MATHS
FOR
MUMS
AND
DADS

About the Authors

ROB EASTAWAY has written several best-selling books that connect maths with everyday life, including *Why do Buses Come in Threes?* and *How Many Socks Make a Pair?* He appears regularly on radio, and has given talks about maths across the UK to audiences of all ages, at locations ranging from the Royal Exchange Theatre to Pentonville Prison. He is married with three children.

MIKE ASKEW is Professor of Maths Education at King's College London. He taught for several years in primary schools in London before moving to work in teacher education. He has published numerous reports, articles and teacher guides, and is constantly in demand, nationally and internationally, for his entertaining and incisive conference talks. He is also a skilled magician.

MATHS
FOR
MUMS
AND
DADS

Rob Eastaway and Mike Askew

◪ SQUARE PEG

Published by Square Peg 2010

6 8 10 9 7

Copyright © Mike Askew and Rob Eastaway 2010

Illustrations copyright © Chris Lyon 2010

Design copyright © Peter Ward for Random House 2010

First published in Great Britain in 2010 by
Square Peg
Random House, 20 Vauxhall Bridge Road,
London SW1V 2SA

www.rbooks.co.uk

Addresses for companies within The Random House Group Limited
can be found at: www.randomhouse.co.uk/offices.htm

The Random House Group Limited Reg. No. 954009

A CIP catalogue record for this book is available from the British Library

ISBN 9780224086356

The Random House Group Limited supports The Forest Stewardship
Council (FSC), the leading international forest certification organisation. All our titles
that are printed on Greenpeace approved FSC certified paper carry the FSC logo. Our
paper procurement policy can be found at www.rbooks.co.uk/environment

Mixed Sources
Product group from well-managed
forests and other controlled sources
www.fsc.org Cert no. TT-COC-2139
© 1996 Forest Stewardship Council
FSC

Printed and bound in Great Britain by
MPG Books, Bodmin, Cornwall

CONTENTS

Preparation

Arithmetic – And How it has Changed

Beyond Arithmetic

The Questions

Answers

And finally . . .

PREPARATION

Q. Find x.

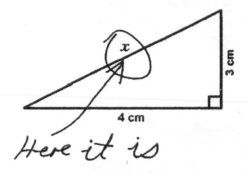

We can't vouch for the authenticity of this famous example of a 'creative' answer to a maths question, but all the others that you'll find in the book genuinely cropped up in the classroom.

INTRODUCTION

It's the moment that many parents dread: 'Can you help me with my maths homework?' Many years, perhaps decades, ago, you went through all this as a child yourself, of course. But it feels very different when you're the one from whom help is being sought. And in any case, things have changed: the maths has changed, the methods have changed, and perhaps the attitudes of children to their parents have changed. That's certainly what parents say, though parents may have been saying this for centuries.

Of all the subjects covered at school, maths is the one that worries parents the most. We've met many a parent whose concern is that their child is beginning to bring home work that the parent can't do. But there are many parents who are at least competent, often very good, at maths, and they have a different worry. The problem is: *they do it differently at school these days*. Attempts by Dad to demonstrate subtraction by adding one and paying ten at the bottom are met by a glazed look from a bewildered child, who finally goes to Mum saying, 'Dad is just confusing me.'

This book is to help you as a parent to re-engage with maths, to see the subject in a new light, to help you to understand *why* they do things differently these days (some of it is for very good reasons), and to help you to better understand what is going through your child's head when she or he claims to 'just not get it'. More than anything, our aim is to put a bit more enjoyment into maths at home – something that seems to be in desperately short supply at the moment.

School maths is a huge subject, and we couldn't hope to cover it all in a single book. We've therefore concentrated on the basics, the stuff that children cover (or should cover) in primary school. We say basic, though much of it is really not that basic at all. In fact, some of the questions that have been posed in tests for eleven-year-olds (commonly known as SATs) would tax most adults. But in this book there will be no sines or cosines, no calculus, and absolutely no quadratic equations. We'll save those for another day.

THE BIG QUESTIONS

Talking to mums and dads, there are some issues that crop up again and again. We've called these the Big Questions. We have identified four of them, and they are so important that we have brought them to the front of the book.

1. Why do they do it differently these days?

When their children reach the age of six, many parents get a dreadful shock. Not only do they discover that what used to be called 'maths lessons' are now called 'numeracy', but they find their children are bringing home vocabulary and methods that the parents don't recognise. In many households, this introduces a problem. Parents keen to help realise that (a) they don't understand what the child is doing, and therefore don't know if it is right or not, and (b) when they try to demonstrate how to do something, all they manage to do is confuse the child. As a result many parents feel frustrated and helpless.

So what is going on, and, as a parent, what can you do about it?

In most schools, the days are gone when maths lessons were endless pages of sums to be done in silence, using techniques for addition and multiplication that dated back hundreds of years. Today's lessons are much more about collaboration and investigation: long periods of silence are rare, even in maths lessons.

The techniques have changed, too. One example of how things have changed is the way that children perform a large multiplication such as 79 × 43. Most parents were taught the method of 'long multiplication', and many still use it to do calculations on the back of an envelope. However, there aren't many parents who can explain *why it works*. It is, if you like, a black box for which you turn the handle and the correct answer comes out at the far end (you hope). Today, the emphasis in schools is on teaching methods that help children to understand the underlying maths, thus (in theory) reducing the chance of them making mistakes, and building a foundation for understanding more complicated mathematics later on.

This move from learning techniques – learning *how* – to understanding the maths behind them – learning *why* – came about for several reasons. Firstly, it was realised that there wasn't a halcyon time when everybody left primary school able to perfectly carry out long multiplication, long division, and so on. Surveys into levels of adult understanding knocked this myth on the head.

Secondly, technology is changing the world we live in at an increasingly rapid rate. Calculators have taken over slide rules and tables of logarithms. The number of times that anyone in their everyday lives actually needs to do a long division or long multiplication has dramatically decreased. But the knowledge of *when* to multiply or divide has increased (which supermarket offer gives me the best value? Which deal on that car is better?), and we need to be confident that the answers provided by a spreadsheet or calculator are reasonable ones. Today, children are taught paper-and-pencil techniques but these are closer to 'back of the envelope' methods that confident mathematicians use.

These techniques are not just about getting the right answer (the calculator is far and away the most sensible tool for that) but about helping children develop insight into mathematics, helping them to develop number sense. As one writer, Richard Skemp, puts it, it's the difference between providing someone with a list of instructions for getting from A to B and providing them with a map. With a list of instructions, making a mistake can send you down the wrong path and it's difficult to get back on track. With a map you can plot your own most sensible route through. Maths teaching these days tries to help children develop mathematical maps rather than remember lists of directions.

As a parent, it is important that you learn about the new techniques. You'll find the most important ones described at various points in this book. The ones you really need to know are the **grid method** for multiplication, the **chunking** or **grouping method** for division and the notion of the **number line** and how it is used. There's also some unfamiliar terminology that you'll want to get familiar with, because children are now having to learn and use terms like **partitioning, array** and **Carroll diagrams**. Use the glossary to find out about these and other particular terms that are causing you concern.

You can at least take comfort from the fact that while some of the techniques being used in school today may have new names, such as the **compensation** method, the techniques themselves are ancient. Indeed, some of them pre-date the techniques that you learnt at school yourself. The Romans and Egyptians used a form of partitioning to add numbers quickly, and now your children are doing it too. Don't be blinded by the jargon that goes with all of these new methods – the principles are very straightforward, and very old.

And the other good news is that, as children progress up the mathematical ladder, all of the techniques eventually join up with the ones that you did at school. Once they are confident in multiplication, for example, your children should be doing long multiplication calculations just like the ones you are used to.

2. How can I overcome my own fear of maths?

If you are one of the lucky parents for whom maths is a breeze, you can skip this question. But you might want to linger for a second, just to discover how some of your peers feel about maths.

For some parents, maths is a living hell. Show them a Year 6 'SAT paper' and they feel physically sick. It's hard to know for sure what proportion of adults have feelings of fear or nausea when exposed to mathematical questions, but on the evidence of our own informal research, it could be as many as 30 per cent (and there are some who need to be reminded that 30 per cent means the same as three out of every ten).

Parents who have a fear of maths are terrified of being found out – by their children. One mum that we spoke to expressed the thoughts of many: 'When my children were at school, I lived in fear of their maths homework. I would do anything I could to avoid coming into contact with it, in case they asked me for help. Fortunately their dad is quite good at maths, so I was usually able to fob them off by saying I was busy, and could they ask Dad instead. But I felt terribly guilty, I felt as though I was letting them down. After a while, they stopped asking me for help with homework – they knew the score.'

Where does this fear come from?

You'll often hear the claim that there are people who don't have the 'maths gene' (which of course suggests that they have also passed those genes on to their children). Does this explain maths-phobia?

The answer to this is almost certainly no, because there is no such thing as a 'maths gene'. How could there be? Humankind has only been doing algebra, probability and calculus for a few hundred years. Most adults today, even those who label themselves as bad at maths, are far more mathematically sophisticated than all but a tiny fraction of adults that lived in medieval times, and it takes thousands or even millions of years for genes to evolve. So whatever it is in our brains that makes some people excel at maths while others struggle, it can't be a gene that is dedicated to maths. (Scientists think that maybe maths ability was a by-product of our need to communicate through language, which itself needs a high level of abstract thinking.)

When you talk to most people about their dislike of maths they more often than not end up talking about a teacher – or a parent – who made them feel inferior. It's not maths itself that people are frightened of but the fear of being shamed.

And it is shocking to discover just how many parents and grandparents have bad memories of maths lessons in the old days (and by old days we're not just talking about the Second World War, we're talking about the 1980s too). Stories of ritual humiliation in front of the class are common. Some recall punishment and physical pain: 'Perkins, what's seven times eight?' 'Er, fifty-four?'. (Perkins then ducks as the board rubber flies over his left ear.)

For some there was psychological pain, too. 'My worst nightmare,' said one mum, 'was Mr Gregory standing up in front of the class and getting us to chant "Maths is FUN, Maths is FUN".' As this mum pointed out, telling people that something is FUN can actually have the opposite effect to the one intended. And let's be realistic. Anything worth learning takes effort and maths is no different. One of the biggest problems we have is the belief that learning should be effortless and fun. As a result children develop an expectation that if they have to put effort into learning maths then they can't be that good at it. In Japan, which has a tradition of high standards in maths, they place much more emphasis on effort than on ability.

Of course, it's hard to verify how common such night-marish incidents were, but perhaps it only takes one moment of humiliation for the whole maths edifice to come tumbling down. Many parents recall a moment when maths suddenly became a brick wall, and further progress seemed futile. This can happen to good mathematicians, too – it's just that they tend to hit their wall at university or beyond. And many mathematicians relish hitting the wall – they see that as a challenge to overcome.

So what can you do to overcome your own fear of maths?

● Recognise that you are probably better at maths than you think. There's a tendency for adults to label the maths that they *can* do (such as identifying patterns, choosing between competing offers in a supermarket, and challenging statistics published by the government) as 'common sense' and labelling everything they *can't* do as 'maths' – so that being bad at maths becomes a self-fulfilling prophecy.

● Most adults assume that maths is a subject that is entirely about *being able to do it* and getting it *right*. We challenge this view. We think that one of the most important aspects of maths is *being stuck* and *getting it wrong*. A maths question isn't called a 'problem' for nothing – the reason why it's called a problem is that it's expected that there will be some difficulty solving it. Getting stuck is an honourable state, and often the best way of dealing with it is to leave the problem for a while – sleeping on it can work wonders.

● Find some spare time when the children have gone to bed, put on some relaxing music, and have a go at the maths questions at the end of the book, starting with the knowledge that there are children who find all of these questions 'impossible'. You'll probably find some of them difficult too, but some of them you'll get quite quickly, and wonder what the problem was. Compare your reaction with those that we received from other parents – you'll discover that the way you think has a lot in common with many others. There is less to fear when you realise that many others are in the same boat.

3. How can I get my child to enjoy maths and be better at it than I was?

We've combined two questions – about enjoyment and ability at maths – because they are closely related. Children will do better at maths if they spend more time having a go at it, and they're more likely to do that if they are enjoying it. Much of their enjoyment and development in maths will depend on the experience they have at home.

One of the most important influences is positive feedback. You should praise your child for the effort they put into the maths, rather than for being 'smart' or 'quick'. The important thing is to help your child develop a 'growth' attitude towards learning maths: they may not be able to do something immediately, but that doesn't mean they will never be able to do it. If children are praised for being generally smart or quick, then when they hit a bit of maths that they cannot immediately do – and they will – then they might believe they have reached their limit and give up.

An ideal time for giving feedback is when you are working through your child's homework with them. When a child gives a wrong answer to a maths question, it's tempting to immediately tell them they have made a mistake and to then explain the right answer. Resist that temptation. Ask them to talk through how they worked it out, and lead them on until (if you're lucky) they spot the error.

To coax more out of them, you can do some of the explaining yourself . . . and if necessary make their mistake in your own workings and then correct yourself with a laugh saying, 'And then three plus three is seven – hang on, that's not right, silly Mum [or Dad] . . .' When your child is explaining give them plenty of time to work through to the end of their explanation. Often the first error that gets revealed is actually the result of some other more fundamental misunderstanding. Letting the full explanation unfold also means that sometimes children realise for themselves where they went wrong. From this, they'll learn that getting it wrong isn't something to be punished for, and even parents get it wrong sometimes.

And when your child gets some maths right, *ask them to explain their working then, too!* This gives you a chance to check their reasoning (sometimes people come up with the right answer for the wrong reasons), but it does something more important, too. If you only ask for an explanation when they get the question wrong, they begin to associate being asked to explain with failure, and so will start to clam up rather than reveal their mistakes. As a parent, you don't have a hope of helping them to reason if they won't reveal their thought processes.

Be tolerant of your child getting stuck. It's very easy when working one-to-one to think, 'I have to get the point across' or, 'Why don't they get it?' Learning does not happen in an instant. Taking a break, coming back to something the following day or leaving it for a week or so can do wonders for understanding – and tempers.

You need to make maths exciting instead of a chore. And above all, NEVER describe yourself as 'hopeless' at maths. This is our biggest 'don't' and we return to it in the 'Dos and Don'ts' chapter at the end of the book. If you show interest in maths, your children will become curious too, and by talking about and playing with maths as a natural part of your daily life rather than it being something that's only done under duress when sitting at the table doing home-work, your child is bound to enjoy it more. But how exactly do you do all that? That's what the rest of this book is about.

4. Why do they (or I) need to know this?

Behind the first three big questions, there is another that looms large for some parents.

When your child bursts into tears over her homework because she doesn't understand how to work out what fraction of a shape has been shaded in, or you look at a test paper about finding prime factors and think, 'I've never needed any of this stuff since I left school', it's only natural to question why you and your children are being put through all this.

It's something that has vexed the education system for many years. As a result, there are endless debates about what should and shouldn't be in the compulsory curriculum. It is a fact that, like it or not, your children *will have to* do this stuff, and that makes many parents cross.

Some maths is very easy to justify because it has obvious applications in life. Any child can see that basic arithmetic will be useful in helping them to work out ingredients for a recipe, calculate change, measure their height or figure out how much they are going to have to save to buy a new game. There is other maths that parents can see will be of benefit even if the children can't yet: percentages, estimation and interpreting statistics will prove essential life skills when the children leave home and start fending for themselves.

The problems begin to arise with maths that is more abstract.

When are they going to need to know about prime numbers? How could knowing the internal angles of a pentagon ever be of practical use to them?

Attempting to answer the question 'What's the point of this?' can be an uphill battle. You can quickly find yourself asking what the point is of anything, if it doesn't have an obvious practical use for you now or in your future life. What's the point of knowing that Henry VIII had six wives, or that magnesium makes a bright white light when you burn it? If you buy the idea that knowledge and learning are useful things, and that it is good to know things for the sake of knowing them, then maths belongs in that foundation of knowledge just as much as Henry VIII. For some parents, that might be enough, but for most there has to be more to it.

One important reason why your child has to do so much maths, whether or not they like it or have an aptitude for it, is that a qualification in maths has become an essential ticket to the majority of professional careers. Whether you actually *need* all that maths is largely irrelevant – what's important is that society has deemed it is a requirement, and having done so it's not likely that society is going to change its mind. So if you want your child to have at least the *option* of being a nurse, a mechanic, a lawyer or a computer game designer, then they are going to have to get a qualification in maths. That's not an argument we like, but it is a fact of life that cannot be ignored.

There are some who justify maths by saying, 'Maths teaches you to think, and how to solve problems creatively.' This is true, but it's far too abstract an idea for most children to take on board, and in any case (some would argue), doesn't a PlayStation do exactly the same thing? Actually the PlayStation argument is not a strong one, as the sort of thinking that most computer games teach you is very limited. A good maths education gives you thinking skills that you can apply everywhere, particularly in developing sound, rigorous reasoning skills, and in being able to see patterns and to make credible predictions. Those skills can be applied to big ideas, such as understanding the shape of the universe, but they are just as applicable in something as mundane as knowing what sort of pension you might expect to receive in thirty years' time.

But the best response to 'What's the point?' is often to answer it with another question: 'Why does there have to be a point?' What's the point of a Sudoku? There isn't one, yet that doesn't stop millions of people doing them, and enjoying it too. Is there a 'point' to reading poetry? While part of the 'point' of maths is to give your children practical skills that they can carry on into adult life, actually the relevance doesn't matter if they are enjoying themselves. And enjoyment doesn't have to mean a laugh a minute. Playing football, climbing a mountain and many other enjoyable pursuits usually feature many moments of discomfort, frustration, even pain, but without them the whole experience would be less rewarding.

And it is the enjoyment factor that is really at the root of those who question the purpose of maths. For most people, it is not their innate ability to do maths that is the problem, it is that they don't enjoy it. Part of the problem is the way that so much maths is presented. If your experience of maths as a child was of tedious, repetitive exercises to practise techniques that other people had already discovered, then it's hardly likely that you came out with a positive impression of the subject.

Play is an essential part of maths, at all levels. That is why games feature so prominently throughout this book. And curiosity is also important. That is why we have also included some challenges that your child may find intriguing. We don't expect everyone to develop an appetite for maths, just as not everyone has an appetite for history or geography. But if children aren't exposed to maths as an intriguing, enjoyable subject, then they are unlikely ever to develop an appetite for it.

MATHS PROPS
FOR MUMS AND DADS

As a mum or dad, most of the opportunities you have to 'talk maths' with your child will be at home. Having certain key everyday items around the house will increase the chance for maths to crop up in conversation spontaneously.

A prominent clock in the kitchen (or wherever you normally have breakfast). If you can have an analogue clock AND a digital clock then so much the better, since comparing and understanding the times on the two clocks becomes an everyday habit.

A traditional wall calendar. Calendars are a good way of getting familiar with counting days, but they also have some subtle patterns in them. One of the columns will be the 7 times table. You can find other patterns by looking at the numbers along diagonals, in square clusters of four, and so on.

Board games that involve dice and spinners. Familiarity with dice and spinners not only helps with counting but also builds an understanding of chance. And spinner games will feature a lot in maths questions at school.

A pack of traditional playing cards – and a few games up your sleeve (such as Snap and Blackjack). Card games are a natural way of learning about sorting and chance.

A calculator. A basic one is enough. This is partly for helping your child when the use of a calculator is expected, but more importantly it's for playing calculator games (see the Calculator Maths chapter).

Measuring jugs with scales. Your child will encounter these in school, so having them at home makes them comfortable with the idea. Jugs that show pints as well as litres provide an instant, visual conversion. Collect empty shampoo bottles or water bottles so children can create their own measuring jugs.

Dried beans, macaroni or Smarties. These are useful for counting large collections to investigate how many are left over if you scoop a large handful and divide them into twos, threes and so on.

A tape measure and ruler. Involve your child when measuring up for furniture, new curtains and DIY. If you make sure you hold the zero end of the tape, they have to do the reading.

A family bar of chocolate (the type that has four rows of eight chunks, for example), stored away in the cupboard for emergency use when it comes to talking about fractions. Chocolate is a great motivator, and good for rewards too.

And some you might want to invest in . . .

Fridge-magnet numbers and symbols. An impromptu way of bringing mathematical equations and questions into the home. We know a dad who, after bedtime, would put calculations like '$7 \times 9 = ?$' on the fridge, and just leave them there as a mystery waiting to be solved when the children come down for breakfast – just imagine the smiles on their faces . . .

Old-fashioned kitchen scales, where the ingredients are balanced by weights. Not only is this a great, tactile way of adding numbers (or fractions if you have old weights), it also introduces the idea of an equation, where the things on one side of the scales are 'equal to' the things on the other side.

A dartboard (with Velcro darts). Darts teaches not only addition and subtraction, but also makes doubling and trebling a familiar exercise. And at the end of a game of darts you are forced to create sums that will fit the target: 'How can I get 47 in two darts, finishing on a double?'

Games with unusual dice. Dice don't have to be cubes. The traditional dice-rolling cricket game 'Howzat' uses hexagonal dice (it's a great travelling game because it is so portable). Other games, particularly fantasy games, use dice with twenty triangular faces (known as icosahedra).

Dominoes. This game seems to by dying out, but you can help to revive it. Dominoes are often used to illustrate combinations (in this case, all the ways of combining the numbers 0 to 6). They are also great for toppling games, when you line them up on their ends and knock the first one over . . .

Guess Who? Children of all ages enjoy this game in which you have to work out which of twenty-four characters your opponent has chosen. The game is a perfect illustration of how to divide things into categories (in this case men and women, people with glasses and people without, and so on).

An indoor/outdoor thermometer. A great device to keep in the kitchen, which tells you what the temperature is indoors and outdoors. In winter the numbers go negative, so your children naturally become accustomed to ideas of freezing, 'below zero' and the negative symbol.

GLOSSARY

By the age of eleven, your child is expected to be familiar with several hundred words or expressions that belong to mathematical vocabulary. Teachers are encouraged to make children familiar with all the words on these lists, and will sometimes even send them home for parents to study.

Hang on, maths vocab tests? What's going on? It may sound like a huge imposition, but in fact most of the terms are so everyday that it would be surprising if they didn't crop up in conversations at home – words like biggest, remainder, quarter-past, reflection, pyramid and litre, for example. Even many 'technical' mathematical terms, such as vertical and horizontal, pentagon, ascending or approximate will be so familiar to you that you won't have any difficulty explaining them to your child if they ask you.

There are, however, some words whose exact meaning you have forgotten, while other terms may even be new to you. We've put those words in this glossary for you. Instead of rigorous mathematical descriptions we've opted for everyday descriptions that, while not precisely right, will do in most situations. If you can't find the word you are looking for in here then either (a) we thought the word was sufficiently commonplace not to need to explain it or (b) it's not a word that primary children are generally expected to know.

Word/ expression	'Everyday' Description	Visual/ numerical example
Acute	A sharp angle (one that is less than 90 degrees)	
Array	Numbers, letters or shapes arranged in a rectangle	An array with three rows and four columns: x x x x x x x x x x x x
Axis	The labelled horizontal or vertical line that marks the edge of a graph	
Axis of symmetry	Same as 'line of symmetry'	
Bar chart	A way of displaying data where bars or columns represent amounts	
Biased	Usually a dice or coin that has been rigged to be unfair. A coin with two-heads is very biased (towards coming up heads!)	
Breadth	Same as width. Although we tend to think of breadth and width as measuring 'side to side' on a rectangular shape, if that side is the longest we usually call it the 'length', and the other shorter side becomes the width (or height).	Usually 'breadth' or 'width' means the *shorter* dimension, i.e., 3 in this example, but there is no formal rule 3cm 12cm

Word/ expression	'Everyday' Description	Visual/ numerical example
Bus-stop method	Alternative name for short division, where the 'dividend' is written inside what looks like a bus-stop	$\begin{array}{r} 8\ 1 \\ \hline 3\ )\ 2\ 4\ 3 \end{array}$
Carroll diagram	A table in which items are sorted under different headings. In most examples it is a table with two categories along the top and two down the side, but you can have as many rows or columns as you like.	Numbers 6–12 sorted into a Carroll diagram: ODD EVEN *One digit* 7, 9 6, 8 *Two digits* 11 10,12
Chunking (or grouping) method of division	A method for dividing by subtracting 'chunks' or groups of the divisor from the dividend	$\begin{array}{r} 7\ \overline{)\ 749} \\ \underline{700} \quad \times 100 \\ 49 \quad \times\ 7 \\ \hline 107 \end{array}$
Circumference	The distance around a circle	
Compensation	A method for adding or subtracting that involves doing a simpler calculation and then making an adjustment at the end to 'compensate' for the answer being different to the expected one	$643 - 498$. Calculating $643 - 500$ is easy: 143. But that is taking away 2 more than 498 so we have to compensate for this by adding 2 onto the answer

Word/ expression	'Everyday' Description	Visual/ numerical example
Compass	A drawing tool used to draw circles (more formally called 'a pair of compasses'). [The device used to navigate, which is circular and also has a point, was named later.]	
Concave	Curving inwards (think of going into a cave)	
Concentric	Shapes or objects that have the same centre (so one will be inside the other)	
Cone	A three-dimensional shape with a round base and curved surface that tapers to a point	An ice-cream or traffic cone
Congruent	If two things are congruent, they are the same size and shape (but might be flipped over or in a different position)	Congruent triangles:
Convex	Curving outwards	
Cube	A cuboid with all the six sides made out of squares all the same size	Stock cubes and dice are cubes. All cubes are also cuboids, but not all cuboids are cubes. All cubes and cuboids are also prisms, but not all prisms are cuboids

Word/ expression	'Everyday' Description	Visual/ numerical example
Cuboid	A three-dimensional shape with six rectangular faces	Cereal and washing powder boxes are usually cuboids. Since all squares are particular types of rectangles, some cuboids have square faces.
Cylinder	A three-dimensional shape with two round faces and one curved surface	Most tins are cylinders
Decimal fraction	Any fraction in which the bottom number is 10, 100, 1,000 or higher power of 10	$\frac{13}{100}$ or 0.13
Denominator	The number on the bottom of a fraction	In the fraction $\frac{3}{4}$ the denominator is 4
Depth	Can mean the distance from top to bottom or from front to back	The depth of water in the fish tank would usually be the distance from the top of the water to the bottom of the tank. The depth of a puppet theatre would usually mean the distance from the front to the back.
Diameter	A straight line joining the edges of a circle that passes through its centre. The same as the width of the circle.	

Word/ expression	'Everyday' Description	Visual/ numerical example
Digit	Any of the numbers 0 to 9	The second digit of 1,398 is 3
Distribution	The pattern formed by a set of data. This pattern is usually most easily seen when the data is plotted in a graph or chart.	The distribution of the heights of children aged 10 120cm 140cm 160cm
Dividend	In a calculation A divided by B, A is known as the dividend	In the sum 39 divided by 3 equals 13, 39 is the dividend
Divisible	Means that a number can be divided exactly with no remainder	44 is divisible by 11, but not by 3
Divisibilty test	A test to check if a number divides exactly without having to do the entire calculation	171 divides exactly by 3 because the digits 171 add up to 9 (a multiple of 3). See page 162.
Divisor	In the calculation A divided by B, B is known as the divisor	In the calculation 39 divided by 3 equals 13, 3 is the divisor
Dodecahedron	A solid shape with twelve faces, each of them a pentagon	
Equals	Meaning two things are the same amount or values, connected with the = sign	7 equals $2 + 5$
Equation	Any mathematical expression, using numbers, letters or symbols, that has an equals sign in it	$2 + 7 = 9$ $A + 3 = B \times 2$

Word/ expression	'Everyday' Description	Visual/ numerical example
Equilateral triangle	A triangle whose three sides are the same length (and whose three angles are each 60 degrees)	
Empty number line	A line used to represent a 'number line'. Children use it to write numbers as an aid to doing sums.	3 20
Face	A flat surface on a three-dimensional object	A cube has six faces A cylinder has faces (the round ends) and one curved surface
Factor	A whole number that divides exactly into another number	3 is a factor of 15. The full set of factors of 15 is 1, 3, 5, 15
Factorise	To find the numbers that, when all multiplied together, make the original number. Normally when factorising you are asked to find the factors that are *prime numbers*	Factorise 35 and you get its factors 5 and 7. Factorise 24 and you get 3 and 8, which further factorises into $3 \times 2 \times 2 \times 2$
Formula	A way of showing how two things are connected, using mathematical symbols. Formulae are nearly always equations (something equals something else)	The formula for working out degrees Fahrenheit from Centigrade is: $F = 1.8 \times C + 32$
Fraction, improper/ vulgar/proper	A vulgar, or common, fraction is any number represented as one whole number divided by another whole number	$\frac{7}{8}$ Proper fraction

Word/ expression	'Everyday' Description	Visual/ numerical example
	The fraction is called 'improper' if the top number (numerator) is larger than its bottom number (denominator), otherwise it is called 'proper'	$\frac{13}{9}$ Improper fraction
Frequency chart	A table recording data and how frequently things occur. Tallies are often used to record frequencies. A survey of traffic passing the school might record the frequencies of buses, bikes and so on, using tallies to note the types of vehicles passing.	
Greater than	Symbol used to indicate when one value is greater than another	$>$
Grid	A rectangle made up of horizontal and vertical lines	
Grid method	A method of teaching multiplication in primary school, as a precursor to teaching the traditional long multiplication method	See page 147
HTU	An acronym for a three-digit number (Hundreds Tens Units)	473
Integer	A whole number (negative whole numbers are also counted as integers)	$0, +1, +2, +3$ and $-1, -2, -3\ldots$
Intersecting	Crossing over or overlapping	

Word/ expression	'Everyday' Description	Visual/ numerical example
Inverse	Doing the opposite or reversing something. If there is more than one step, then the inverse is to start at the final step and work backwards.	The inverse of $+2 \times 5$ is $\div 5 - 2$ The inverse of 3 up, 2 right is 2 left, 3 down
Inverse operations	Addition and subtraction are inverse operations. Multiplication and division are inverse operations.	
Icosahedron	A solid shape with twenty faces, each of them a triangle	
Isosceles triangle	A triangle in which two of the three sides (and therefore two angles) are the same	
Kite	A four-sided shape, with two pairs of adjacent sides that are the same length (usually the shape of a classic kite)	concave kite convex kite
Length	When describing a shape or object, the length is most commonly used to describe the longest dimension	
Line of symmetry	Any line (real or imaginary) that divides a shape into two parts, where one part is the mirror image of the other part	

Word/ expression	'Everyday' Description	Visual/ numerical example
Lowest common multiple	If you have two whole numbers, their lowest common multiple is the smallest number into which both of the numbers can exactly divide. The lowest common multiple can also be found for more than two whole numbers.	The lowest common multiple of 6 and 15 is 30; the lowest common multiple of 2, 6, 7 and 15 is 210
Mean	The most common form of average, calculated by finding the total of a group of numbers and dividing by the number in the group	For the numbers: 1, 3, 3, 5, 7, 8, 8 the mean is $1 + 3 + 3 + 5 + 7 + 8 + 8$ divided by 7 (the number of numbers in the string)
Median	One way of expressing the average – the median is the middle value if all the items are listed in order from smallest to largest	The median of 1, 3, 3, 5, 7, 8, 8 is 5, the middle number. If there is an even number of numbers in the list, the median is found by taking the mean of the middle two numbers: the median of 1, 3, 5, 6 is 4 i.e. $(3 + 5) \div 2$
Minus	The more formal way of saying 'take away'. Minus is also (confusingly) often used as an adjective to describe negative numbers.	10 minus 7 equals 3

The temperature is minus 7 |

Word/ expression	'Everyday' Description	Visual/ numerical example
Mixed number	A number expressed as a whole number and a fraction	$1\frac{1}{4}$ (one and a quarter)
Mode	An alternative way of expressing an average, the mode is the measurement that occurs most frequently	The modal number of fingers on a human being is 10. (The mean is slightly less than 10 because there are many people with fewer than 10 fingers.)
Multiple	The result of multiplying a whole number by another whole number	28 is a multiple of 7 because 7×4 is 28
Multiplication bond	The multiplication facts in the times tables (now usually up to 10×10)	$4 \times 9 = 36$ $2 \times 8 = 16$
Negative	A number that is below zero	
Net	A flat sheet that can be folded up into a three-dimensional shape	The most common net used to make a cube:
Number bonds	All pairs of numbers that add to a particular number you are interested in	The number bonds of 10 include 2 and 8, 4 and 6. Number bonds of 13 include 4 and 9.
Number line	A line with counting numbers written underneath it, used to help children to learn addition and subtraction	
Obtuse	A 'blunt' angle, an angle that is more than 90° and less than 180°	

Word/ expression	'Everyday' Description	Visual/ numerical example
Octahedron	A diamond-like solid object made up of eight triangles	
Operation	The four operations in primary maths are addition, subtraction, multiplication and division	
Parallel	Two lines that never meet, most easily remembered if you think of the word as para | | el	The lines forming the top and bottom of this shape are parallel:
Partitioning	Breaking a number up into smaller numbers, usually hundreds, tens and units (typically done to make addition or multiplication easier)	146 can be partitioned in many ways, for example 100 + 46 Or 100 + 40 + 6 Or 120 + 12 + 14
Perimeter	The distance around the edge of a shape, or the line that marks that edge	The perimeter of the rectangle is indicated by the thick black line 2m 3m The perimeter is 10 metres
Perpendicular	Two lines that are at right angles to each other	

Word/ expression	'Everyday' Description	Visual/ numerical example
Pictogram	A graph that uses pictures to represent several objects. For example a stick drawing of a person might be used to represent ten people	= 10 people
Pie chart	A circular chart used to display the relative proportions of a group that had a particular feature	
Place value	The value of a digit determined by its place in a number	The place value of 2 in 526 is 20, the place value of 2 in 246 is 200
Polygon	A two-dimensional shape having three or more straight sides	
Positive	Any number that is larger than zero	
Prime numbers	The whole numbers that are not divisible by any other number other than 1 and the number itself	The first few primes are 2, 3, 5, 7, 11, 13... (1 is not a prime)
Prism	A three-dimensional object whose two ends are the same polygon, and which has the same cross-section along its length	
Product	The result of multiplying two or more numbers together	The product of 7 and 3 is 21. The product of 0.5 and 0.5 is 0.25
Protractor	A device used to measure angles	

Word/ expression	'Everyday' Description	Visual/ numerical example
Pyramid	A three-dimensional shape with a polygon as a base and triangular faces that taper to a point	
Quadrant	One region in a shape that is divided into four quarters through its centre	
Quadrilateral	A shape with four straight edges	
Quotient	The answer in a division sum	In the sum 39 divided by 3 equals 13, 13 is the quotient
Radius	The distance from the centre of a circle to its edge	
Random	An event where there is more than one possible outcome, and where the outcome cannot be predicted	The number that appears on top when you roll a dice is random
Ratio	The relative amount of one thing compared to another, usually written using a colon	When diluting fruit cordial, the ratio of water to cordial is 5:1
Reflective symmetry	A fancy way of saying that half of a shape is a mirror image of the other half	
Rhombus	A four-sided shape in which each side is the same length. A square is a special type of rhombus.	

Word/ expression	'Everyday' Description	Visual/ numerical example
Right angle	When the angle between two straight lines that meet is 90 degrees. The angle is indicated by a square.	
Right-angled triangle	A triangle with one angle that is 90 degrees	
Scalene triangle	A triangle whose sides all have different lengths	
Set square	Usually a flat right-angled triangle made of plastic that is used to help you draw perpendicular lines	
Sign change	When you change a number from positive to negative (or vice versa). This appears as a button on most calculators.	Sign change $+2$ makes it -2
Sphere	A three-dimensional object that is circular whichever direction you look at it	
Square metre	Used as a measurement of area. (Be careful to say 'square metres' and not 'metres square'. If a carpet is '12 metres square', it is 12 metres long and 12 metres wide, i.e., its area is 144 square metres.)	A carpet that is 4 metres long and 3 metres wide has an area of $4 \times 3 = 12$ square metres
Square root	If you have a number A, then its square root is the number B such that $B \times B = A$. The symbol $\sqrt{}$ means 'find the square root of'.	$\sqrt{25} = 5$ (Strictly speaking -5 is also a square root of 25, since $-5 \times -5 = 25$)

Word/ expression	'Everyday' Description	Visual/ numerical example
Squared	A number multiplied by itself (so-called because the area of a square is the length of its side multiplied by itself). Usually written with a little raised number 2 (known as the 'index').	$5^2 = 25$ x^2 is spoken as 'x squared'
Standard unit	The normal units of measurement for length, weight, temperature and so on. These can be imperial or metric.	foot metre litre kilogram second
Sum	In everyday language a sum means any calculation, for example, the sum 6×7. But strictly speaking it should only be used to refer to addition calculations.	
Symmetry	The mathematical term for what we intuitively recognise; images that have a reflective or rotational pattern to them; butterflies, faces, the letter S, etc.	S
Tetrahedron	A pyramid with a triangular base and three triangular faces	
Trapezium	A four-sided shape with exactly two sides that are parallel	

Word/ expression	'Everyday' Description	Visual/ numerical example
Units	The final digit of a whole number (the numbers 0 to 9)	7,814 ← Units are 4 23 + 14 The units 37 ← column
U.t	Acronym for a number that has one digit after the decimal point (Units.tenths)	7.4
Venn diagram	A diagram that allows you to group items into two or more categories, some of which might belong to more than one category	
Vertex	The mathematical name for a corner	A square has 4 vertices, a cube has 8
Volume	How much space there is inside an object, measured in litres or cubic metres (or larger/smaller scales of these measures)	This cube is 2cm × 2cm × 2cm so its volume is 8 cubic centimetres
Width	There is sometimes confusion about which dimension of a shape is its width. See *Breadth*.	

A YEAR-BY-YEAR SUMMARY

Here's a brief summary of the maths that your child is *likely* to encounter as (s)he progresses through primary school. For years, primary schools have followed a national framework that was set up by the government. This provided a model of which topics should be covered at which stage in which year. Some schools still follow this so religiously that you can even predict what particular tasks your child is going to be given in week 3, Year 4.

However, this framework is only a guideline, and your child may be going to a school that approaches the subject at a different pace, and in a different order. And, of course, schools do have the option of including different material as 'enrichment'. For example, the national curriculum for primary children does not include anything about 'pi' (the ratio of a circle's circumference to its diameter) but many young children will encounter it, perhaps through an enthusiastic teacher, and if your child is ready for such ideas then it does no harm to introduce them before the timetable says they need to know it. (In fact, we think pi is an extremely *good* thing to know about!)

This is why we have deliberately kept this section brief. We don't want you to become obsessed by whether your child is ahead of, or behind, the 'right level' because this can inadvertently lead to you putting unwanted pressure on your child. ('Sam, the book says you should be able to do pie charts by now, what's the matter with you?')

Before going through the Years in detail, here is a quick summary of two other bits of terminology that confuse some parents:

Key Stages 1, 2, 3 . . .

When the government introduced the National Curriculum they also introduced numbers for all the years of a child's schooling. Children start formal schooling in the year that they turn five, and start in Reception. After reception, they enter Year 1 (Y1), followed by Year 2 (Y2): these two years of schooling are known as Key Stage 1 (KS1). Key Stage 2 (KS2) starts with Year 3 and children leave KS2 at the end of Y6. Key Stage 3 is the first part of secondary schools, Y7, 8 and 9 and KS4 is the GCSE years, Y10 and Y11. And KS5 is Y12 and Y13, what used to be called Sixth Form.

Levels

Just to confuse matters, along with naming the Key Stages and the Years, the National Curriculum introduced the idea of *Attainment Levels*, commonly referred to as 'Levels'. These are used to describe children's progress. You might think it would be sensible to match up levels with years but, instead, each level describes, roughly, two years' progression (we say roughly because some children progress more rapidly in some years). Most children are expected to reach Level 2 at the end of Year 2, and the majority are expected to reach Level 4 at the end of Year 6. At the time of writing between 75 and 80 per cent of children leaving primary school achieve Level 4, but that does not mean that 25 per cent of children leaving primary school have 'failed' or are 'innumerate'. Children with Level 3 and even Level 2 still know a lot of mathematics. An aside (as this is a book about maths not literacy) you only need to be operating at L2 to be able to read the *Sun* or *Mirror*!

Reception

Counting is important and the basis of arithmetic. Children in Reception will be learning to say the number names in order, forwards and backwards, and to count collections of objects. Their teachers will be encouraging them to use the language of more and less and to compare numbers.

In Reception children work on questions like 'what is 1 more than 6?' or 'can you say the number 1 less than 7?' involving numbers up to 10. They engage in activities such as putting two groups of everyday objects together and finding the total. They may remove some objects from the group and so be introduced to subtraction as 'taking away'.

The beginnings of multiplication and division are developed through counting groups of the same size, for example, finding the total number of sweets if there are three sweets each on four plates. They might share nine biscuits between the three bears as an introduction to division.

Making patterns, building models and sorting things around the classroom develops their mathematical reasoning although at this stage they will be using everyday language to describe these. They will compare things and use language like 'bigger', 'greater', 'heavier' and 'lighter'.

Year 1

In Year 1 the counting that children began in Reception is extended to counting collections up to 20 and recording the total. Finding and saying what is one more or one less than a number is still important, now with numbers to 20.

In Y1 children learn pairs of numbers (the number-bonds) such as, 4 + 6, 7 + 3 and the other combinations that make 10 exactly. They learn that addition can be reversed, for example that 2 + 8 is the same as 8 + 2. Although it may appear a big step, they also figure out calculations like adding 30 to 56 or subtracting 20 from 63 by using informal written methods or practical materials.

Skills needed for multiplication are developed by learning to count in twos and fives and learning the doubles of 1 to 10. They share collections of objects into equal groups to explore division.

Y1 children also make patterns, pictures and models using common 2-D and 3-D shapes and name the shapes they have used. They will also talk about the position of things using everyday language like 'behind', 'above', 'next to'. Estimating, measuring, comparing and weighing objects help them understand measuring. Talking about when things happen in the day or putting pictures of events in order introduces them to talking about time. Block graphs and pictograms help them display information.

Talking about, say, half an apple, or a quarter of the children introduces the language of fractions.

Year 2

In Year 2 children's knowledge of reading and writing numbers is extended up to 1,000, although they are only expected to be able to order numbers that are less than 100. They learn about odd and even numbers.

Now number bonds like $5 + 2$ or $8 - 3$ are learnt and the patterns in counting in tens help children to answer calculations like $50 + 20$ or $80 - 30$. They work on mental methods to do calculations, such as $36 + 40$, $45 - 8$, that is adding or subtracting single digits or multiples of 10. A key idea that they will be getting to grips with is that subtraction is the inverse of addition: knowing that $16 + 7 = 23$ means you also know that $23 - 7 = 16$.

In multiplication they will be doubling numbers to 20 and halving the answers, and the tables are introduced, starting with the 2, 5 and 10 times tables.

In shape and measures they learn names like square, cube, sphere and other common two- and three-dimensional shapes. They look at symmetry in pictures and 2-D shapes. Making and talking about quarter, half and full turns begins their journey into the world of angles. Their measuring starts to be more accurate by using metres, centimetres, kilograms and litres and they learn to read divisions on scales. They gather data on topics like different ways to travel to school and represent it in tables, diagrams, block graphs and pictograms.

Year 3

If your child is in Year 3 they are likely to be working with whole numbers up to at least 1,000 and learning about where to position numbers on a number line. Counting on or back in tens is a core skill and they learn that numbers like 365 can be split (partitioned) into 3 hundreds, 6 tens and 5 ones, and that 365 rounded to the nearest 10 is 370 and rounded to the nearest 100 is 400.

Now they are learning to add mentally pairs of numbers like $63 + 37$ or $48 + 52$ and other combinations that make 100 exactly. When they are confident with this they will mentally figure out calculations like $23 + 43$ or $36 - 21$. Written methods will be used to explain addition and subtraction of two-digit and three-digit numbers that are not 'friendly' enough to be done mentally.

The 3, 4 and 6 times tables will be the ones they are learning, as well as how to recognise multiples of 2, 5 and 10 up to 100, such as 465 being a multiple of 5. Calculations like 48×10 or 35×100 help them to explore how multiplying by 10 or 100 works. Another big idea is introduced: division is the inverse of multiplication. This helps them work out division calculations from multiplication facts that they know: knowing that $6 \times 9 = 54$ means you also know that $54 \div 9 = 6$.

Proper fractions are introduced and diagrams used to compare them. They figure out simple fractions of numbers and amounts, for example $\frac{1}{2}$, $\frac{1}{3}$ or $\frac{1}{4}$ of 12 kg

Work on angles is extended to recognising and naming right angles. They work on the relationships in measures, for example how kilometres and metres are related and they learn to read scales. Telling the time to the nearest 5 minutes on a clock with hands is a skill they develop. They use Venn and Carroll diagrams to sort information.

Year 4

Now children move from working with whole numbers and meet decimals (up to tenths and hundredths), particularly in relation to money and measurement.

Mental calculations like $700 + 600$ or $6,000 - 3,000$ continue to build on their knowledge of bonds ($7 + 6$, $6 - 3$) and they work mentally with less friendly calculations involving two-digit whole numbers such as $36 + 28$ or $81 - 14$. They start to use written methods to add and subtract two-digit and three-digit whole numbers and money calculations involving pounds and pence.

In multiplication they will be doubling two-digit numbers, multiples of 10 and multiples of 100, for example, double 28, double 80, double 600. Knowing your tables is extended up to 10×10 and they mentally carry out multiplications such as 357×10 or $480 \times 1,000$. Written methods will be introduced for multiplying and dividing two-digit numbers by a one-digit number.

The idea of equivalent fractions is introduced using diagrams and now they put fractions as well as whole numbers on a number line. They identify pairs of fractions that make 1 and carry out calculations using fractions, for example finding $\frac{1}{5}$ of 30 apples or shading in $\frac{5}{8}$ of a rectangle.

Y4 children work on the ideas of vertical and horizontal and measure angles in degrees. They find areas and perimeters of rectangular shapes. Now they tell the time to the nearest minute using different clock notations, for example a.m., p.m. or the 24-hour clock.

Year 5

By Year 5 children are mentally adding, subtracting and doubling simple decimals, for example, $6.5 - 2.7$ or double 2.7. They use written methods to add and subtract large whole numbers and decimals with up to two places, such as $23.45 - 17.67$.

Mentally they multiply numbers like 300×500, that is, pairs of multiples of 10 and 100. They find factor pairs of two-digit numbers, for example the factor pairs of 36 include 3, 12 or 4, 9. They learn about common multiples and find them for pairs of small numbers, such as 6 and 4 having 24 and 48 as common multiples. Other mental calculation methods include things like learning to multiply by 25 by multiplying by 100 and dividing by 4. Written methods are now used to multiply and divide three-figure numbers (Hundreds Tens Units, or 'HTU') by single-digit numbers. They also work on multiplying two-digit numbers together, and multiplying numbers with a decimal, for example 8.3×4.

As well as solving fraction problems like finding $\frac{1}{100}$ of 5 litres they find percentages of numbers and quantities such as 10% and 15% of £60.

In Y5 children use co-ordinates and recognise and construct parallel and perpendicular lines. Measuring becomes increasingly accurate, for example to the nearest millimetre. They meet the idea of mode as a measure of average.

Year 6

In Year 6 children find the difference between positive and negative numbers in a context, for example working out how much the temperature must drop from $+5^{\circ}C$ degrees to $-4^{\circ}C$.

By Year 6 children are using written methods to add and subtract integers (positive and negative numbers) and decimals.

They use their knowledge of the tables facts to work out mentally decimal multiplications and divisions, for example, 0.6×7 and $3.6 \div 9$. They figure out squares of multiples of 10 such as 60×60. Prime numbers less than 100 are explored and the prime factors of two-digit numbers, for example factorising 24 into $2 \times 2 \times 2 \times 3$. They learn to multiply and divide integers and decimals by a one-digit number and do multiplications like 352×73.

Y6 children relate fractions to multiplication and division, for example $9 \div 3 = \frac{1}{3}$ of $9 = 9 \times \frac{1}{3}$. They find fractions and percentages of whole number quantities such as 55% of £240.

They calculate angles in a triangle and convert between metric units using decimals, for example converting 3.45 litres to 3,450 millilitres and vice versa. They use mode, range, median and mean.

Into Year 7 and beyond (secondary school)

By now children work on ordering fractions by converting them into decimals and they use ratio notation. They are familiar with the ideas of multiples, factors, divisors, common factors, highest common factors and lowest common multiples. They calculate percentage increases or decreases and calculate efficiently.

They learn to calculate the area of right-angled triangles and volume and surface area of cubes and cuboids.

They work with the probability scale from 0 to 1 and carry out statistical inquiries.

And then, of course, come the delights of trigonometry, algebra and much, much more.

■ ■ ■ ■ ■ ■

So that's an overview. One thing you'll discover (if you haven't done already) is that topics get revisited again and again, not just through primary but also into secondary. Partly this enables children to revise what they have learnt, but also at each stage the idea is to introduce another layer of information, to build this edifice called a maths education. So it's not a question of, 'Age 7 – learn division – OK, sorted . . . next?'.

In any case, it is the school's job to provide the structured learning, not yours. Your role is to nurture and support your children's maths knowledge away from school, to bring it into their real lives, and, most important of all, to help to turn it into an exciting adventure. In the following chapters, we've not tied most of the maths to particular year groups, since many of the ideas and games can be used with a very wide range of ages and abilities. And to be honest, there's stuff that Year 3s are taught that (shh, say it quietly) *many mums and dads find challenging.*

ARITHMETIC –
AND HOW IT HAS
CHANGED

NUMBERS
AND PLACE VALUE

Q. What is half of 8?

A. 3

Because 8 is to
3's put toghether.

Children's first encounter with maths is naming the numbers and learning to count, something that many children master by the time they start school. So if your children are already way past that stage, you might be inclined to skip this chapter. Before you do, however, you might want to pause just to be reminded of what an ingenious and sophisticated thing our number system is: an ancient Roman or Greek teleported to a Year 1 classroom would be in awe of this system of numerals in which a '1' might represent one thing, or ten, or even a thousand. And as for decimal points, and our quaint system of giving multiples of ten names like 'thirty', they would have been baffled. Numbers aren't nearly as easy as we adults (who have got used to them) like to think, and it should be no surprise that many children struggle with understanding 'place value' for many years after they learn to count.

In this chapter, we'll give some background to how we ended up with the counting system that we have, an overview of how schools now teach it and some games and activities that reinforce the number system, which can be as enjoyable for ten-year-olds as they are for infants.

Common problems children have with numbers and place value

1. Thinking that 6,000 is one more than 5,099
2. Writing 'one hundred and thirty-six' as 10,036
3. Not appreciating that there are 24 groups of 10 in 243, not just four 10s.
4. Thinking that 3.453 must be smaller than 3.35 because it has thousandths in it.

5. Thinking that 0.75 is smaller than 0.203 because 75 is smaller than 203.

The story of working in tens

Our system of counting works by grouping things together in tens, putting tens of tens together to make hundreds, tens of hundreds to make thousands and so on. The choice of working in tens is, of course, down to the fact that we have ten fingers (or digits) on our hands – and ten on our feet, too, just for good measure.

So familiar is this system of tens ('deci-mals') to us, it seems almost natural – it's just how numbers are. But this idea of place value as we know it, with Hundreds, Tens, Units and the rest, and its use daily with money and measures, is only a few hundred years old. Just as it takes children many years to learn to 'decode' written symbols and become proficient readers, so it takes an equally long time for them to become proficient with our invented system of saying, reading and writing numbers.

The idea of grouping in tens was well established many centuries ago, in the use of the abacus. Early abaci were made from clay with grooves that could hold up to nine small pebbles. Once ten was reached the ten pebbles would be replaced by one pebble, placed in the next groove. When that was filled with nine pebbles, the tenth would result in all ten again being replaced by one in the next groove. There was no need to record any calculations – the pebbles helped you keep track. Different symbols were used to represent the larger groups, in the Roman system, for example, X for ten and C for 100.

These early counting systems had no symbol for zero. After all, if a column on the abacus had no pebbles in it, then there was no need to record the empty space. To the Romans, 305 was simply CCCV — that there were no tens was obvious from the fact that no Xs were recorded.

THE ROMAN NUMBERS

I	1
V	5
X	10
L	50
C	100
D	500
M	1,000

Roman numbering was based entirely around the use of seven letters. You'll notice that these were for 1, 10, 100 and 1,000 (the numbers we use as the places in our decimal system) but also the half values of 5, 50 and 500. For numbers of 4,000 and above, they put a horizontal bar above a number to represent 'thousand', so for example $\overline{X}$ represents 10,000. In a Roman number, I, X and C don't always mean 1, 10 and 100. When placed to the left of (respectively) X, C and M they mean *subtract* 1, *subtract* 10 and *subtract* 100. So for example IX means $10 - 1 = 9$, and CD means $500 - 100 = 400$. The most common use of Roman numbers today is as a traditional way of indicating dates (requiring some quick deciphering when they roll up the credits at the end of a TV programme).

Test yourself

i) On which London monument would you find the Roman number MDCLXVI (each of the letters, in descending order), and why?

How place value was invented

The adoption of our modern (Arabic) numeral system began to change things. The same symbol – for example 3 – was used to represent 3 units, or 3 tens, 3 hundreds or millions and so on. Now what mattered was the position of the symbols as well as the symbols themselves. So when there were three pebbles in the hundreds groove on the abacus and five in the units and no tens, people started to record 3 5. But what about that space between the 3 and the 5? How could it be made clear that the gap where the tens would normally go was deliberate, rather than someone's messy writing? Or that only the tens were missing and not hundreds as well and that the 3 really stood for 3,000? The problem was solved by the invention of zero – 0 – as a place holder. Putting zero between the 3 and 5 – 305 – held these digits in their rightful places of units and hundreds. Now the value of the 3 is not ambiguous: its value has to be 300. And hence the name of our system – place value.

Recording numbers using place value really took off with the invention of the printing press. As paper became cheap people could begin to set aside their 'old' calculating technology – the abacus – and take up a new, more versatile technology – pen and paper. Some historians suggest that there was as much debate then over the 'dumbing down' impact of this new pen-and-paper technology as there is today over the pen-and-paper versus calculators debate. Presumably 'what will you do if your quill breaks?' was the equivalent of today's 'what will you do if the batteries go flat?' argument. In fact, there are parts of the world where, to this day, the abacus is still the preferred calculating tool. In Japan they use a particular form of abacus – the sarabon – and proficient users are quicker and more accurate than using pen-and-paper.

To explore further our taken-for-granted place value system, let's think about how things might have turned out if we only had 8 fingers instead of ten.

What if we didn't have ten fingers?

Our counting system works by counting on our fingers until we run out. When we have used up all of our fingers, we need to start again, so to record that we have twelve items, we say one full set of fingers, plus two more – which we write as 12. This is a big step for children – to relate the 1 in 12 to '1 group of ten'. To help you to get into the mindset of your child and appreciate the challenge this presents them, it is helpful to become a novice again and to work with an unfamiliar number system. So imagine what maths would be like if we didn't have ten fingers. What if humans only had eight fingers (like Bart Simpson or Mickey Mouse)? Our counting would go: 1, 2, 3, 4, 5, 6, 7, 10, 11, 12 . . .

This is known as working in *Base* Eight. Notice how this counting system never actually uses the numeral 8. In this system, 10 does not stand for ten, it stands for one group of eight and no units. So – in the world of eight fingers, 12 is one group of eight and two units, which means ten in our usual system.

Test yourself

ii) Can you figure out what 124 in base 8 stands for in our base ten system?

You can extend this idea of bases to any number of fingers you like. Imagine an alien with only two fingers. It would never use the numeral 2. Instead, it would count the first three numbers: 1, 10, 11 . . . and after that? Two-fingered maths never uses the number 2, so after 11 comes 100. Then 101, 110, 111, 1,000 (so 1,000 represents the number 8, one eight, no fours, no twos and no units). Working with two fingers is known as Base Two, more often referred to as the Binary system.

Many parents – and certainly grandparents – studied numbers in different bases at school. There was a good reason for this, in that every day they were likely to encounter counting systems based on numbers other than 10. For example there were 12 pennies in a shilling and 12 inches in a foot, 16 ounces in a pound, 8 pints in a gallon. There is more about this in the chapter on Measuring.

Now that most of the world is metric, there is a much less obvious need to learn about numbers in other bases, but an understanding of how they work gives a useful insight into the decimal system that we take for granted. And for anyone wanting to understand the fundamentals of how computers work, an understanding of binary numbers is vital (you'll find more on pages 241 and 297).

Game: Twenty

This counting game is extremely popular in playgrounds around the country. Children as young as five can play it, but it will also amuse teenagers and even adults. It comes in many variations, but the basic game is called 'Twenty'. Two players take it in turns to count up from 1 to 20, counting one, two or three numbers in their turn (each player can decide how many to count during the turn). The player who ends up having to say 20 loses. So a game might go like this:

Ali: One, two.
Jake: Three.
Ali: Four, five, six.
Jake: Seven, eight.
Ali: Nine, ten, eleven.
Jake: Twelve, thirteen, fourteen.
Ali: Fifteen sixteen . . .
Jake: (smiles as excitement grows) seventeen, eighteen, nineteen!
Ali: Drat – twenty.

Children have been know to play this game multiple times, trying to figure out what the strategy is. It quickly dawns on them that getting to nineteen is the key, since then their opponent has no choice but to say twenty. But how do you guarantee to get to nineteen? The answer is to make sure you get to 15. After 15, whatever your opponent says (16 or 16, 17 or 16, 17, 18) you can get to 19 in your next move.

In fact, there turns out to be a pattern – to win the game you need to get to the 'stepping stones' 3, 7, 11, 15 and 19:

1 2③4 5 6⑦8 9 10⑪12 13 14⑮16 17 18⑲20

To guarantee to win the game, you need to count up to 3. This is easy if you go first – you just say one, two, three. If you go second you have to hope that your opponent doesn't count to 3 on their first turn. You then count up to 7, 11, 15 and 19 in your other turns.

This may seem to be a game about counting, but in fact it goes much deeper than that, since it's really also about discovering patterns.

You can immediately make the game a new challenge simply by changing the rules. For example, what if you change the target to 25? What if you allow players to count up to *four* numbers? What if there are three players?

The challenge of counting in groups

Before looking in a bit more detail at our system for counting large numbers, let's spend a moment considering the complexity of what young children have to get to grips with in learning to count.

Obviously they have to learn the counting words: one, two, three and so on. And although it is obvious to adults, they have to learn that these words have to be said in the same order – unlike naming the teddies on the bed. All the traditional nursery rhymes – One, two, buckle my shoe; One, two, three, four, five, once I caught a fish alive – all developed to help children with counting.

Children's early experience of numbers are usually as

'adjectives' rather than as 'quantities': 'I'm four' is no different for the child from 'I'm Sally' or 'I'm a boy'. House numbers, keys on mobile phones, television channels – children are surrounded by numbers as labels and they do not see them as having anything to do with quantities. Of course, as parents, we help them start to make this connection, but there is still a lot to do. For example, if you put out six sweets, count them with a four-year-old and then ask the child to give you three, children will go through a stage of handing over the one sweet that was pointed at when the word 'three' was said, rather than picking up three altogether. It is a big step from going 'one, two, three, four, five, six', pointing to the last sweet when you say six and realising that 'six' can now be a label for the whole collection of sweets, not just the last one pointed to. And that is assuming that the child has got over the hurdle of co-ordinating three things: pointing to a sweet at exactly the same time as saying the next number word, making sure you don't skip over either a sweet or a number word, and making sure you don't count the same sweet twice. Learning to count is best done by counting real objects rather than pictures – objects can be moved aside as they are counted, so they don't get counted twice and the movement can be co-ordinated with the saying of the number name.

Now imagine you are six or seven and just beginning to feel confident with this game called counting and you can please everyone by showing one finger and counting until you've put out ten fingers. Then someone comes along and starts calling your ten fingers 'one'! Life was so much easier for Roman children, who had their different symbol, X, to represent this.

The same difficultly also shows up when children start to learn about money. Why should a five-pence coin, given that it is smaller than a single penny, be worth the same as five individual pennies? Give me the five pennies any day, thinks the young child.

Playing games and activities that involve collecting, grouping, exchanging and naming can help here. Make up simple dice games that involve collecting pennies and every time you have five pennies change them for one five-pence coin. Or collect coloured counters. Every time you have ten red counters, change them for a blue counter. If you get ten blue counters change them for a yellow one. Talk about the counters; 'Look, I've got two blue counters and three red ones. If I changed them all for red counters, how many would I have?'

QUICK TIP

Look out for opportunities to talk about groups of things as single quantities while out and about. Talk about multi-buys in the supermarket. 'If we buy two six-packs of cola, how many single cans is that? The orange juice is in four-packs. We need twelve cartons for the week. How many four-packs shall we buy?'

The eccentric naming of numbers

A second major obstacle is our language for counting in place value. Frankly, it's a mess! We say sixteen and seventeen, but eleven and twelve rather than one-teen or two-teen. We say sixty-five (six—ty five) and seventy-eight (seven-ty eight) so why not two-ty three instead of twenty- three or three-ty six instead of thirty six? The Chinese are much more logical. Their name for 13 is one-ten three and 36 is three-ten six. Adding 16 + 13 is much easier if you say 'one-ten six plus one-ten three must be two-ten nine' rather than 'sixteen plus thirteen is twenty-nine'. Some have argued that the head-start that children who learn Chinese counting get in understanding basic numbers explains why countries in the Far East consistently out-perform Western countries in maths tests.

At least the English can be thankful not to be lumbered with the quirky French language, which, after counting reasonably logically in tens up to 60 (soixante) suddenly changes tack to describe 70 as 'sixty ten', 80 as 'four twenties' and 90 as 'four twenties ten'.

In English it gets worse when we write things down. We say 'sixty-seven' and write 67 – logical enough, the order of the digits matches the order in which we hear them. But we say seventeen yet write 17 – the order of the digits is the reverse of the sound order (whereas for the same number the French logically say 'ten-seven'). Children who write 61 for sixteen are not simply getting it wrong, they are sensibly trying to link what they hear and say with how to write it. And eleven and twelve provide no clue as to what to write down at all.

Further traps await with the numbers in the hundreds. Seventy-three has two words – seventy and three – and two symbols to record it: 73. But four hundred and seventy-three sounds different. Saying the word 'hundred' can lead children to recording 40,073 rather then 473.

Game: Nice or nasty

'Nice or nasty' is a great game to play for creating and reading three- or four-digit numbers. You need: A pack of playing cards (remove the picture cards), paper and pencil. An Ace counts as the number 1.

Each player draws three boxes side-by-side on a piece of paper, each box big enough to hold a playing card.

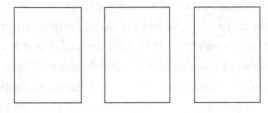

Decide whether you are trying to make the biggest or smallest number. Shuffle the cards and place them in a face-down pile. Take it in turns to turn over the top card and place it in one of your boxes. When each player has placed three cards, read these out as a three-digit number – 2, 5, 9 is two hundred and fifty-nine. The winner is the player with the largest or smallest number as previously agreed.

That's the nice version of the game. The 'nasty' version is when you choose to play a card either on your own board or on your opponent's. If the whole family plays and you are aiming for, say, the first player to win five rounds, a lot of strategy comes into play as you decide who you might try to prevent from winning a round. You can also play the game with four-digit numbers.

The tactics are interesting. If you are playing the nice version and aiming for the largest number, and you turn over a 1, where do you put it? Since it's a low number you ought to put it at the right. But what if you draw a middling number like five? Tricky – do you play safe and make this your first digit, or do you gamble on getting a bigger card next time? It's like a junior version of Bruce Forsyth's old game show *Play Your Cards Right*.

Play with large numbers

Given that we are stuck with our naming system, what else can you do to help? It is helpful to playfully explore larger numbers with your child. It's tempting to think that the logical order to work with numbers with children is from small to large: numbers to 10, numbers to 20, numbers to 30 and so on. But it is precisely with these smaller numbers that our naming is such a mess. Playing with bigger numbers helps children realise that there is an overall logic and that the whole system is not chaotic. As you are walking along, suggest counting together and play with counting through suggesting things like: 'Let's count and start at sixty.' 'Let's count and start at seventy-five and let's take it in turns to say the next number.' Adopt a playful attitude. 'Let's count in tens and start at forty. Fifty, sixty, seventy, eighty. What's going to come next? Ninety.' Be delighted if your child suggests that ten-ty should come after nine-ty; it most logically should.

Playing with really big numbers also helps. We group the digits of big numbers in threes, and that's how we say them, too. Think about how you say this number: 876,452,781. (Sometimes spaces are used instead of commas in large numbers. This is to avoid confusion with the French system, where the comma is often used instead of the full stop to represent the decimal point.)

Listen to the pattern: eight hundred and seventy-six *million*, four hundred and fifty-two *thousand*, seven hundred and eighty-one. The rhythm of 3 starts to become clear. Hundreds, tens and ones are the basic building blocks for the naming of all larger numbers.

Game: Race to 100

Another place value game. You need: a dice, paper and pencil.

The aim of the game is to score exactly 100 before your opponent does.

Take it in turns to roll the dice. You can either take the score as shown, or take ten times that value. So a roll of 4 can count as 4 or 40. Keep a running total of your scores – the first person to reach 100 exactly is the winner. You cannot go over 100.

Mike			Rob		
Roll	Score	Running total	Roll	Score	Running total
3	30	30	5	50	50
4	40	70	6	6	56
4	4	74	2	20	76
5	5	79	1	10	86
2	20	99	4	4	90
3	0	99	2	2	92
1	1	100 Mike wins			

A variation on the game is to start with 100 and subtract scores.

Inside kids' heads

Here is a typical question that your child might be asked to test their understanding of place value. Write out the following numbers in order, largest first:

901   1,001   910   99   109   190   999

And here is the answer from one child:

$$999 \quad 99 \quad 910 \quad 901 \quad 190 \quad 109 \quad 1001$$

Can you see why she came up with this answer?

This child is focusing on the digits in the numbers rather than the value of digits as determined by their place in the number – she's thinking, 'All those nines must make 999 bigger than 1,001'

It can help your children if, while they are getting used to place value, you write out numbers in columns with headers. The numbers 1,001 and 999 are written out as:

Th(ousands)	H(undreds)	T(ens)	U(nits)
1	0	0	1
	9	9	9

Even and odd

Once children are confident with counting, they can begin to discover patterns in numbers. One of the simplest patterns is even and odd numbers, which children as young as four can begin to appreciate. Walk down most streets, and you can point out that houses don't 'count' in the normal way. On one side of the street they go 2, 4, 6, 8, 10 . . . , while on the other side they go 1, 3, 5, 7, 9 . . . Children will quickly pick up the pattern, so that you can ask 'what's the next house number going to be' and they will count on by two.

You can make sure that the pattern of even numbers sticks with them by adapting the popular football chant that counts in twos: 'Two Four Six Eight – who do we appreciate? EVEN NUMBERS! And don't forget ZERO.' It takes all of ten seconds to teach this chant to a child, and they'll readily chant it back to you again . . . and again . . . and again. They particularly seem to like the 'don't forget ZERO' bit at the end, if you do it in a sing-song voice.

Game: Even-Odd Cards

Find five plain cards, the size of postcards. Using a black pen, write 0, 2, 4, 6 and 8 on the five cards. Then using a red pen write 1 on the back of the 0 card, 3 on the back of 2, and so on with 5, 7, 9 on the other three cards. Deal out the five cards into five piles on the table, then turn your back, and ask your child to turn over as many of the cards as he wants – he can turn over just one, or all five, or anything in between, and he can choose which cards he can see. There will now be some red and some black cards face up on the table, and you have no idea how many of each.

You now announce that even though you have no idea which numbers are facing up, (because your back is still turned) you will be able to work out what the numbers add up to. All he needs to tell you is how many red numbers he can see. Suppose he says he can see two red numbers. You pretend to be doing some complicated formula about red and black, adding and subtracting and then announce that the total is . . . 22. You turn around, and with your child add up the numbers, and sure enough they add up to 22.

The secret is extremely simple. However many red numbers your child says, you simply add that number to 20 to get the answer.

Why does it work? Suppose all the black numbers are showing. These add up to 20. If you turn over a black card, you increase the total by one, because the red (odd) number on the other side is one higher. It doesn't matter which cards are turned over, if there are, say, three red cards showing then the total must be 20 + 3.

Counting down

Your child has truly mastered counting when they can count backwards as well as forwards. Countdowns before a rocket takes off are a godsend for imprinting reverse counting in children's minds (an old *Thunderbirds* video will do this for you: FIVE . . . FOUR . . . THREE . . . TWO . . . ONE . . . THUNDERBIRDS ARE GO!)

There's also a little backwards counting trick that has been known to entertain children who already know that they're supposed to have ten fingers.

'Did you know that I've got eleven fingers?'

Child looks curious and studies hands. 'No, you've got ten!'

'I'm going to *show* you that I've got eleven. First there is my left hand with . . . [point to fingers on left hand in turn] ten, nine, eight, seven, six fingers [hold little finger when saying six]. So that's six. Then add the five on the other hand, six and five makes eleven'.

Your temporarily baffled child will then get very excited demonstrating that you've actually got ten.

Game: Magic colour prediction

This little game is great for learning counting, but it makes a great trick for older children too – for them the challenge is to figure out how it works. Secretly write down the colour orange on a piece of paper and place the prediction face down where everyone can see it. Show your child the diagram below, which looks like a large 9. Think of it as a circle of colours, with a tail going off at the bottom.

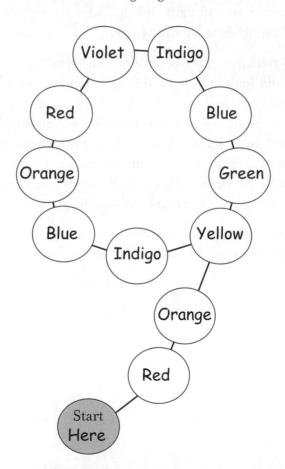

1. Start with your finger at the end of the tail, on the 'Start Here' circle.
2. Ask your child to choose a number bigger than two but smaller than ten.
3. Count her number going up the colours of the rainbow, so 1 is Red, 2 is Orange, 3 is yellow, 4 is Green, etc., stopping on her number.
4. Now count her number again, this time going clockwise and around the circle instead of down the tail.

You will always end on the colour orange.

It will work for numbers bigger than 10, too, though for numbers bigger than 11, you find yourself doing more than one loop of the circle. Choosing a big number, like 50, and counting it both ways very quickly out loud – and ending up on orange yet again – makes great entertainment.

This sort of game has a double value. Not only is it embedding the idea of counting, but there's the deeper question of *why it always works*. This is problem-solving at its most enjoyable.

The biggest number you can think of

What's the biggest number you can think of? 18 . . . 94 . . . a
hundred . . . a thousand. What IS the biggest number? Is
there a biggest number? It's a game that children love to
play.

It helps that the big numbers all start to sound a bit silly.
Millions, billions, trillions . . . squillions! It's another good
way of talking about place value, because the number of
zeroes is very important.

Imagining a million of anything is extremely hard. Even
a thousand seems vast to most children. You can stimulate
their imagination by picking up on a big number quoted in
the news – the transfer fee of a footballer, for example.
'Ronaldo has just been sold for £80 million', the news might
say. So you ask: 'Hey, how long will it take to buy Ronaldo if
you start saving your pocket money – if your pocket money
is, say, £10 a week?'

You might be surprised to learn that most children
reckon a year's worth of pocket money should do it. Wrong!
Ten years? A hundred years? They ask, with increasing
excitement.

The correct answer is actually 160,000 years. Even that
number means little to most children. So you should
explain that 160,000 years ago was before the start of the last
Ice Age. Neanderthal man was still roaming the corners of
Europe. Just imagine that caveman thinking: 'You know
what, I fancy having that Ronaldo in my team in 160,000
years' time, so if I start putting my ten mammoth skins
aside every week, I'll be able to buy him.' It does make you
wonder if these footballers can ever be worth it . . .

What's the difference between a million and a billion?

Go to the largest room in your house. Now imagine that the whole width of a room represents a billion. Where along the wall would a million be? Since a million is a big number, it's natural to think that a million would be a fair way along the wall. But if your room is, say, 5 metres wide, then a million would be just 5 millimetres away from the wall. Compared to a billion, a million is tiny. Compared to a trillion, a billion is tiny. Millions, billions, trillions fill our newspapers, they all sound vast, but it's good to develop in your children the idea that there's a vast distance between big, very big and ginormous.

Million	1 000 000
Billion	1 000 000 000
Trillion	1 000 000 000 000
Quadrillion	1 000 000 000 000 000

. . . and so on.

At the ultimate end of ginormous, we get to 'Infinity' . . . until somebody says: 'Infinity plus one.' But what *is* infinity plus one? If you want to know, look at page 307.

The decimal point

Just as numbers can get infinitely bigger, they can get infinitely smaller too.

We've already seen how our counting system works by grouping numbers into tens, with each column in a number being ten times bigger than the one to the right of it (hundreds are ten times ten, thousands ten times hundreds and so on). This pattern can be done in reverse. Reading from left to right, each column becomes ten times smaller (hundreds are ten times smaller than thousands, ones are ten times smaller than tens). But why stop there?

We can divide 'ones' into pieces that are ten times smaller: tenths. And divide those 'tenths' into pieces ten times smaller again: hundredths. We call these smaller numbers decimals. It is connected to the word 'decimate', which originally meant 'to make a tenth of'. (The Romans had a cruel punishment called decimation, which involved randomly killing one in ten of a group of soldiers if the cohort had done something wrong.)

When mathematicians had the idea of decimals, one problem was how to record these new numbers. We could just write $93 \frac{5}{10}$, $\frac{8}{100}$ but someone had the bright idea of using just a dot to record where the whole numbers ended and the decimal part began: 93.58.

And the decimal places can go on for as long as you want:

Units	tenths	hundredths	thousandths . . .
3	5	8	4

Just as numbers get infinitely bigger, they can get infinitely smaller too.

Inside kids' heads

There are two particular situations with decimal numbers that are worth looking out for with your child.

The first results from over-generalising a pattern that children notice about whole numbers. Think about comparing 11,111 and 9,999. Children learn that even though 11,111 looks smaller than 9,999 because of all the ones, it is actually the larger number because it is a five-digit number and 9,999 is only a four-digit number: the more digits a number has the larger it has to be, regardless of the actual digits. Offered two salaries: a four-digit salary or a three-digit salary, we know the four-digit salary is going to pay more without knowing the actual salaries.

Now, children learn that decimal fractions get smaller as the number of decimal places increases: 0.03 is smaller that 0.3 and 0.003 is smaller still. The over-generalising occurs when children think that just as the more digits a whole number has the *bigger* it must be, that the more decimal places a number has automatically makes a number smaller: that 0.125 must be smaller than 0.8 because 0.125 has thousandths in it while 0.8 has only tenths in it. (Note how the language can help cause this confusion: a number with thousands in it would actually be larger than a number with only tens in it and thousandths and tenths sound very similar to thousands and tens.)

You can help your child here by talking about the value of the digits in each number – 0.8 has 8 tenths in it while 0.125 only has 1 tenth in it – without looking at the other digits.

The other over-generalising that children make is to read the digits after a decimal point in the same way as they read whole numbers: reading, say, 0.125 as 'nought point one hundred and twenty-five' makes it sound bigger than 0.85 – 'nought point eighty-five'. Unfortunately, our decimal money system encourages this: we read £3.25 as 'three pounds twenty-five' rather than 'three point two five'. When you come across decimal numbers in contexts other than money, make a point of reading them digit-by-digit rather than in the way we talk about money.

ADDITION
AND SUBTRACTION:
MENTAL METHODS

Harry saves **50p** coins.

He has saved **£8.00**

How many **coins** has he saved? 16

Show how you worked it out in the box.

Addition and subtraction are two of the foundation stones of maths, and the first where mums and dads are likely to encounter unfamiliar language and techniques, such as *number lines* and *number bonds*.

Perhaps the biggest change in the way that children are taught addition and subtraction is that today children are taught mental techniques first, before going on to paper-and-pencil methods. This chapter explains why there has been this change of emphasis.

Of the two topics, it is subtraction that causes children the most problems. Most parents tend to think of subtraction as simply the opposite of addition, but in fact it is more complicated than that because subtraction has so many different meanings. Subtraction can be taking away, finding the difference, even adding. For example, if you have 201 conkers and take away 196, what are you left with? A child probably thinks of this as a hard take-away sum, whereas an adult is more likely to treat it as addition (how many do I need to add to 196 to make 201 . . . easy, five!). It is because addition and subtraction are often the same thing that we have combined them in this chapter and the one that follows. This first chapter looks at why mental methods have become important, what makes them different from mental arithmetic and how you can support your child in developing them. In the next chapter we look at addition and subtraction of numbers that cannot be easily handled mentally and the paper-and-pencil techniques that children are commonly taught these days.

Common problems children have with mental addition and subtraction

1. Counting on or back in ones to add or subtract when there are simpler methods, for example adding on 9 to 17 by counting on, 18, 19, 20, ... when it is quicker to add on ten and subtract one.
2. Using paper-and-pencil methods to calculate when a moment's thought would reveal a simple mental method, for example $245 + 299$ or $4,003 - 2,996$.
3. Thinking that subtraction is only 'taking away' and not realising it can be used for 'finding the difference', for example, how much taller am I than my brother?
4. Thinking that you cannot take a bigger number from a smaller number, so a calculation like $7 - 11$ is impossible (thank goodness the bank doesn't believe this).

Puzzle: Gauss's ingenious shortcut

Once upon a time, or so the story goes, there was an eight-year-old boy called Carl Friedrich Gauss. His teacher was mean, and wanted the children to work hard while he sat and got on with other things. So the teacher asked the children to do a huge sum:

$$1 + 2 + 3 + 4 + 5 + \ldots \text{ all the way to } 100$$

'That should take them most of the lesson,' thought the teacher, expecting the children to have to write out a great number of paper-and-pencil calculations to find the overall total. But within a minute, Gauss had his hand up. 'Sir, I have the answer.' He had found a most ingenious solution: we'll reveal what this is at the end of the chapter. (If you want to mull on how he might have found the answer so quickly, here's a hint: suppose you added all these numbers up twice?)

Gauss went on to become a famous mathematician and we are not suggesting that many children will be as mathematically precocious as Gauss. We do, however, expect children to ask themselves the question 'is there a quick and efficient way of doing this calculation' and not just carrying out the 'traditional' method. Many addition and subtraction calculations, even with large awkward numbers, can be done more quickly, and often more accurately, using mental methods.

Mental or paper and pencil?

An eight-year-old boy was asked: 'If a baby was born in 1998, how old was it on its birthday in 2001?' He unhesitatingly answered, 'Three.'

The same boy was later given the calculation 2,001 − 1,998. Suddenly his common sense understanding of the question was lost and he went into autopilot. Here is his calculation:

$$
\begin{array}{r}
2,001 \\
-1,998 \\
\hline
1,997
\end{array}
$$

He has found the difference between the numbers in each column and written it down. A problem with vertical calculations is that they focus on the digits that make up the numbers, rather than the numbers themselves. This means that children don't think about the calculation in terms of sensible answers.

Mental maths is not intended to be like the mental arithmetic of the 1950s and '60s. Then there was pressure to answer quick-fire questions around the class and often a sense of shame if you weren't as quick as the rest. (One of us, Mike, went to a school where the teachers thought the solution to this was a rap across the knuckles with a ruler. The logic of this is still lost on him.)

Mental maths is about looking at the numbers involved in a calculation and thinking about a sensible calculation method. Before doing any calculation, you should encourage your child to ask themselves, 'Could I do this in my head?'

For example: $2,734 + 3,562$.

The numbers in this calculation are not very 'friendly' and reaching for paper and pencil and setting this out in columns actually is a sensible choice.

But what about: $3,998 + 4,997$.

At first sight, this looks pretty much like the first calculation. But a moment's pause before re-writing this in column form might mean noticing that both numbers are close to multiples of 1,000: 3,998 is close to 4,000 and 4,997 is close to 5,000. Now $4,000 + 5,000$ is easy – 9,000. All we need to do is a bit of adjusting; this is 5 too many, 2 from the 3,998 and 3 from the 4,997. So the answer must be 8,995. Although written out like this, it sounds long-winded, it's actually quick to do. Quicker than reaching for paper and pencil. And less prone to error.

Test yourself

i) *Mental or pencil and paper?*

Which of these calculations can easily be figured out mentally? Which would you need paper and pencil for?

a. 152 + 148
b. 300 − 148
c. 843 − 677
d. 843 − 698
e. 4,997 + 5,003
f. 6,002 − 3,999

Early addition and number bonds

Before we look at how to help your child develop mental methods like the ones above, it is helpful to look at the early stages of learning to add or subtract.

As adults, most of us learnt addition and subtraction so long ago that we have forgotten how long it took us to get to grips with it. Your average four-year-old will be able to tell you that two bananas plus three bananas is five bananas, or that two splats plus three splats is five splats, even if they don't know what a splat is. But ask them what is 3 + 2 and they will look at you blankly — it is too abstract. (In a classic piece of research, one four-year-old when asked what 3 plus 2 is replied, 'I don't know, I don't go to school yet.')

Once children start school they do learn to add and subtract in the abstract: 3 + 5, 7 − 4 and so on. This is learning the **number bonds**, as your child or their teacher may call them. This is the start of their journey into mental arithmetic, something that features far more prominently in primary school maths today than it did in the past.

Games are a perfect way of giving children a chance to develop their number bonds. Any board game with two dice will be randomly testing their ability to add up two numbers between 1 and 6. You can easily adapt classic games like Snakes and Ladders to play with two dice. A good starting point is to adapt one dice by covering up the usual number with sticky labels and marking the dice with only 1 dot on each of 3 sides and 2 dots on each of the other three sides. Rolling a normal dice and this half-spot dice provides great practice in adding one or two to a number – key skills for the young child.

Dominoes are also good for developing number bonds; give your child a 'quick peek' at a domino – long enough to register the number of dots on each half, but not long enough for them to count them all. Can they figure out the total number of dots? Turn the domino over to check.

The number line

One of the surprising results from research into how children add or subtract is the discovery that using paper and pencil helps develop mental methods! Not using it for traditional vertical calculations, but for child-friendly 'back of the envelope' approaches. Using paper and pencil to jot down what's going on in your head and to keep track of intermediate steps will help your child talk about what they are doing and that then helps them remember the mental methods.

Addition is a natural development of counting, and there is a key moment when your child realises that to add up five sweets and four sweets, there is no need to count all nine – she can start at five, and count on four from there. In school, they'll use a number line to do this, with an arrow starting at five and adding four to make nine.

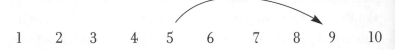

1   2   3   4   5   6   7   8   9   10

As the numbers being added get bigger, it can help to break the addition into steps. For example, instead of adding '8 + 7' children can first to jump 2 (to make 10), then jump the remaining 5, to make 15.

1   2   3   4   5   6   7   8   9   10   11   12   13   14   15

Breaking a number up into parts goes by the grand name of *partitioning*. (It's an idea that most parents are already familiar with – it's just that we never used to have a name for it!) As you will see later, partitioning is useful throughout arithmetic, so it's a handy word to add to your vocabulary.

The simplest subtraction calculations are done exactly the same way, except moving to the left instead of the right on the number line. So the calculation 12 − 5 can be done as a single hop to the left, or in two steps like this, for example:

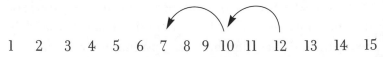

1   2   3   4   5   6   7   8   9   10   11   12   13   14   15

Bigger additions with empty number lines

Once children have learnt the basic number bonds (the result of adding and subtracting numbers between 1 and 10), they are ready to develop their mental arithmetic further. These days, children are encouraged to imagine an empty number line, onto which they will be putting the numbers in a calculation.

Through the number line, children begin to discover that there are many different methods they can use for doing a sum. Let's look at an example:

$$55 + 37$$

Method 1: split both numbers into tens and ones

If your child has developed some confidence in working mentally and a basic understanding of place value, then they might invent for themselves a method of answering this based on splitting both numbers into their tens and ones:

- add 50 and 30 to make 80
- add 5 back on to make 85
- add the 7 to make 92

You can put these steps on an empty number line:

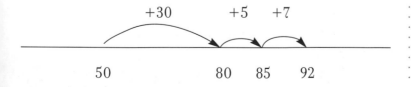

Method 2: split the smaller number into tens and ones

A slightly more advanced approach is to split just one of the numbers into tens and ones. This removes a step from the calculation:

- add 55 and 30 to make 85
- 85 add 7 is 92

The empty number line makes this clear:

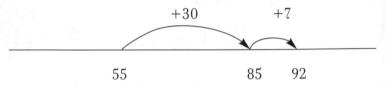

Children seem to lean 'naturally' towards the first method and it is tempting to let them work with what they are most comfortable with. But if you can encourage them to use the second method, it can help when they start working on subtractions.

Test yourself

ii) *Number line*

Use the number line method to do the following sum: 48 + 36.

See how your answer compares to one done by a nine-year-old child.

Practising mental addition

As well as developing these mental methods, your child will still need to practise using them so that they become second nature. You may be wondering about those calculation workbooks that you can buy in the high street, or calculations that you can download from the internet. Is there a place for these? Yes, but talk with your child about different calculation methods. Don't just set them off to work through a page of twenty calculations using the same method for each one simply because they are set out as vertical calculations. You could ask your child to draw a circle around each calculation they think they could figure out mentally, share your favourite ones with them, treat the page of calculations as something to be approached intelligently rather than slavishly.

But games and puzzles are much more enjoyable ways of practising. Here are two.

Game: Cricket, dice and adding

Any parent who has been a 'scorer' at a cricket match will know that it's an excellent game for practising simple arithmetic. During the summer months, you can take your child down to the local club match, and the players will probably be only too happy to let your child help to operate the scoreboard – and children do love being scorers. The team begins its innings at zero and every time it scores your child will be adding a number between 1 and 6. Towards the end of the game, they'll be dealing with numbers in the low hundreds — 'What's 198 plus 4, Dad?'

If you'd rather stay at home, there's a simple dice game called 'Howzat' that is a good simulation of a cricket match. It uses two dice (traditionally hexagonal-cylinder dice, but any dice will do). The first dice is rolled to tell you how many runs were scored (1, 2, 3, 4 or 6). If number 5 comes up, that represents 'Howzat', and you roll the other dice to decide if you are out (1, 2, 3 or 4) or not out (5 or 6). If you are not out, you go back to rolling the runs-dice and adding to your score. Any player who gets to 50 gets a big round of applause. Children have been known to keep themselves occupied for hours playing this game. (Or is one of the authors just looking back at his childhood through rose-tinted specs?)

Game: Addition with a magic twist

Here is a magic grid of numbers that your child will love to investigate. Find yourself a pen and paper and copy the grid onto it:

7	5	6	4
4	2	3	1
6	4	5	3
8	6	7	5

Choose any number in the grid and put a circle around it.

Then cross out all the other numbers in the same row and column as your chosen number. (For example, you could choose the 2 and cross out the other lines in its row and column like this – but choose any number you want!)

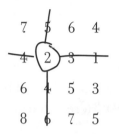

Now circle another number in the grid, and again cross out all the uncrossed numbers in its row and column. Do this for a third number, and finally, put a circle around the only remaining number in the grid. You now have four circled numbers that you chose yourself. Add up the four circled numbers. Is your total 19?

Make a magic number grid

Here's how to make a magic grid that always adds up to 19. It's a great excuse to practise lots of addition sums.

1. Draw a four-by-four grid.
2. Use a pencil to write down eight numbers that add up to the magic number, in this case 19, along the top and down the left-hand side of the grid. It's good if most of the numbers are different, but they don't have to be. For example you can choose these eight numbers (check they add to 19):

	3	1	2	0
4				
1				
3				
5				

3. Now fill in each square in the grid by adding the number at the top of the grid to the number at the side. For example the top left number in the grid is 3 + 4, so fill in 7 (this is known as an addition square). Here is the grid partly filled in:

	3	1	2	0
4	7	5	6	4
1	4	2	3	
3				
5				

4. Rub out the original numbers round the edge of the grid. The magic square is now ready to use.

Now as long as you choose exactly one number from each row and column, the numbers will always add to 19.

To make a magic square that adds up to a different number, 43 for example, just make sure that the numbers around the edge of the grid in Step 2 above add up to 43. Magic number grids make a great birthday card. Put a grid onto the front of the card, and make sure it adds to the age of the recipient.

Subtractions with empty number lines

Subtraction using number lines works exactly the same way as addition, but moving to the left on the number line instead of the right. However, some approaches with subtraction turn out to be easier than others.

For example, look at the sum: $55 - 37 = \boxed{}$

Splitting both numbers into tens and ones, and trying to work mentally can be muddling, unless you are confident with negative numbers. If you aren't careful, it can go like this:

- Subtract 30 from 50 to make 20
- Subtract 7 from 5 (negative 2). Can I do that? Maybe I have to subtract 5 from 7 to make 2
- Do I add the 2 to 20, subtract it, or what? HELP!!!

Things are much more straightforward if you start with the whole first number (in this case 55) and subtract first the tens and then the ones.

- Subtract 30 from 55 to make 25
- 25 subtract 7 is 18

The logic behind this method is again nicely demonstrated on the empty number line.

$55 - 37 = 18$

Test yourself

iii) *More number lines*

Do this subtraction, then compare your answer with a nine-year-old's: $73 - 28 =$

Linking addition and subtraction

Mental methods rely on children being able to mentally add or subtract small numbers to or from larger ones. Two methods are useful, they work equally well with addition or subtraction.

The first method is called *bridging through ten*. If you want to add, say, 6 to 137, then it's helpful to think of splitting the 6: first add 3 to 137 then 'bounce off' 140 and then add on the other 3. Here's a child showing how this works on an empty number line:

$$137 + 6 = 143$$

This works similarly for subtraction.

$142 - 8 =$
$142 - 2 = 140$ and $140 - 6 = 134$

The second method is called *compensating* – you add or subtract more or less than you need to and then 'compensate' for this by adjusting the answer. For example, when adding 9, a compensating method is to add 10 and take off 1:

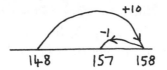

Again, this works for subtraction. $267 - 48$. It's easier to do $267 - 50$, that's 217. But that's taking off 2 too many (you only want to subtract 48 not 50), so the answer needs adjusting by adding 2 back on: 219

Once your child is confident with showing their methods using an empty number line, they may begin to invent other methods that work with particular types of numbers. For example: $55 + 39 =$

You could add on 30 and then add on 9, but confidence with an empty number line means you might do something like add on 40 and subtract 1.

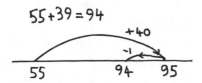

Understanding subtraction: taking away or finding the difference?

Your child will not only be learning how to calculate the answers to addition and subtraction calculations, they will also be learning when to use each operation. Children do not usually have much difficulty in knowing when to add, but subtraction is more complicated.

Many people read a calculation like 37−19 as 'thirty-seven take away nineteen' − their earliest experiences of working with subtraction are as take-away. Count out 37 counters, now take away 19, how many are left? But lots of situations that are not 'take-aways' can be solved with subtraction calculations.

> I have 37 stickers and my friend has 19.
> How many more stickers do I have?

You can solve this by calculating 37 − 19, but nothing has been taken away − at the end of the day, I still have 37 stickers and my friend still has 19.

In the same way, suppose the new Wii game that I want costs £37. So far I've saved £19 in my piggybank. How much more do I need to save?

Children will solve problems like this by counting up from 19 to 37, which can be written as 19 + ☐ = 37, but you can turn this around and say, 'What's 37 − 19?'

The empty number line is again a powerful image for helping children develop mental methods for subtraction and explore the different meanings of subtraction. Think about these three calculations: have a go at figuring out the answers and thinking about how you did this before reading on:

$$130 - 17$$
$$130 - 118$$
$$130 - 49$$

Most people do the first calculation as a 'take-away'− they remove 17 from 130, possibly by taking off 10 to 120 and then taking off 7 to get 113. In contrast, 'taking away' 118 from 130 is cumbersome. It can be done but it's more likely that you say to yourself, 'Well, 12 onto 118 is 130.' In other words, you add on to 118 rather than do a 'take-away' and what you are doing is finding the difference between the two numbers. 130 − 49 sometimes provokes a different 'compensation' strategy − 49 is close to 50, so take off 50 from 130, that's 80, and add one back (compensate for the extra one that was subtracted when taking away 50 rather than 49). All of these methods can be given a visual representation on an empty number line.

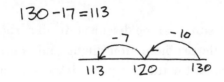

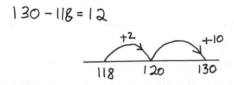

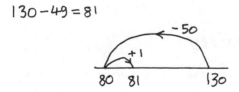

The reason so many teachers now encourage children to make 'jottings' using an empty number line is that there is strong evidence from research in psychology that children begin to work with some sort of mental representation of a number line and can figure out addition and subtraction calculations without writing anything down.

QUICK TIP

Work with your children to read subtraction calculations in lots of different ways. Presented with the sum
10 − 7 you can say: 10 take away 7, 10 minus 7, 10
subtract 7, what is the difference between 10 and 7, how
much greater than 7 is 10, how much less than 10 is 7.

Test yourself

iv) Subtraction sums can be disguised in all sorts of forms.
Here's a real-life problem. You might find yourself using
different methods for the two parts, involving subtraction
and addition:

Rachel buys a pair of sandals for £13.75 and a pair of
trainers for £32.40.
a. How much change does she get from £50?
b. How much more did the trainers cost than the sandals?

Game: Diffy

Draw a very large square on a piece of paper. Ask your child
to choose interesting numbers to put at the four corners.
Mark the mid-point of each side of the square and work
with your child to figure out the difference between the pair
of numbers at neighbouring corners and to write the value
of the difference at the mid-point mark. (Here the difference between the top corner numbers is 8 and the left-side
numbers differ by 11. You can finish it off . . .)

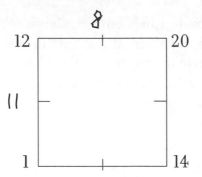

Now join up the four mid-points to make a smaller, slanted square within the one you started with. Mark the mid-points and, using the values at the corners of this slanted square, write in the differences at the midpoints.(So the difference between 8 and 11 is 3.) Join these to make another smaller square, mark the mid-points and keep going.

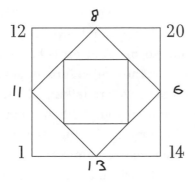

Eventually all the values of the mid-points will settle down to be the same and so the last square you can draw will have zero at each of the mid-points. (The great thing about this is even if you make a mistake somewhere along the line and calculate a difference incorrectly, eventually it will still settle down to four zeros.)

You can turn the game into a challenge. Who can choose four numbers less than 20 that lead to the most squares before you get completely *zeroed*? What happens with bigger numbers? Negative numbers? Fractions? Suppose you don't start with a square but a triangle, or a hexagon? It's a great way to get children doing a lot of subtraction calculations without presenting them with a page of 'sums'.

Developing subtraction further – negative numbers

Another trap awaiting children in subtraction is negative numbers. Part of the difficulty is that negative numbers are much harder to picture than positive ('counting') numbers.

Your child may not formally be introduced to negative numbers until Year 5 or 6, but they are likely to encounter them before then. Many buildings have floors that are underground, so below ground level the lift might go to floor −1, −2, and so on. And on cold winter days, your children will spot that the temperature might be −5 degrees Celsius. In both cases, we are talking about numbers that are 'below' zero, and so you can help them to appreciate those numbers by turning the number line so that it is vertical, so that the negative numbers really are *below* zero. Instead of the normal horizontal number line, you can call it a number ladder. Zero is in the middle, you go *up* for the positive numbers, and *down* for the negative numbers. You can then refer to negative numbers as 'underground' numbers. (The chances are your children won't have encountered overdrafts yet, but of course these, too, are underground numbers.)

Since debts will one day be only too familiar a part of your children's lives, you can even introduce them to the idea of sums involving negative numbers. A plus sign means move up, and negative means move down. So 3 − 5 would mean start at level 3 and move down 5, which takes you to underground 2.

Perhaps not surprisingly, children (and adults) find a sum like 9 − 5 much easier than 5 − 9. But both lend themselves to imagining moving to the left on a horizontal number line or down on a vertical one. Starting at 9 and moving down 5 floors will land you at 4. Starting at 5 and moving down 9 can be broken into two steps: moving down 5 brings you to zero, and another 4 down and you're at −4. (In fact, it's always true that if you reverse a subtraction you get the same answer but negative. So 48 −23 is 25, and 23 − 48 is −25.)

One other confusion with negative numbers is the use of the word 'minus'. Minus can be used as a verb (as in 'eight minus three equals five') or as an adjective ('the temperature was minus six degrees'). The best way to remove this confusion is never to use minus as an adjective. Always refer to the number −2 as 'negative two' . . . or even 'underground two'.

Gauss's ingenious shortcut:

So what was Gauss's trick? Remember the challenge was to calculate 1 + 2 + 3 + 4 + 5 + , . . . all the way to 100.

Gauss's ingenious insight was that instead of adding the numbers in sequence, he wondered, what would happen if he wrote out the sequence 1 to 100 forwards, and then underneath wrote it out again in reverse?

Forwards $1 + 2 + 3 + 4 + 5 + 6 + \ldots + 100$

Backwards $100 + 99 + 98 + 97 + 96 + 95 + \ldots + 1$

Now, Gauss realised, if you add pairs of numbers in the columns together you get this:

$$101 + 101 + 101 + 101 + 101 + 101 + \ldots + 101$$

This is just 101×100, an easy sum with the answer 10,100. The original sum is half of this, in other words 5,050.

This is proof, if any were needed, that sometimes it pays to think about the sum first before ploughing into a standard method. How about that as a bedtime story?

ADDITION AND SUBTRACTION: PAPER-AND-PENCIL METHODS

Q. What is the difference between 9 and 4?

A. The 9 is curvy but the 4 is all strait lines.

In the previous chapter we looked at how children are being supported to develop mental methods for carrying out addition and subtraction calculations. We argued that there will still be times when they are presented with calculations that do not nicely lend themselves to mental methods. The educational world is somewhat split on what are best: paper and pencil or calculators. Some argue strongly that just as paper and pencil was the 'technology' of its day, so children should use today's technology – calculators. Others still argue for paper and pencil on the basis of 'what if the batteries are gone in the calculator' or 'suppose you don't have a calculator to hand'. Given that many children are more likely to have an electronic hand-held device to hand than paper and pencil, it is logically hard to sustain the argument for the priority of paper and pencils (and contrary to popular belief, research shows that sensible use of calculators is not detrimental to understanding). But wheels grind slowly in education and paper-and-pencil methods look set to be around for a good while yet.

Common problems children have with paper-and-pencil addition and subtraction

1. Adding or subtracting numbers digit by digit and not thinking about whether the answer makes sense.
2. Mis-applying half-remembered rules like 'you can't take 9 away from 7' and ending up with 42 as the answer to 67 − 29.
3. Using paper-and-pencil methods when one of the 'mental' methods would be quicker.

Vertical addition – the parents' method

You might be hankering for a reminder of how this addition used to be done. Many of us learnt that adding 146 and 879 is done like this:

$$
\begin{array}{r}
1\ 4\ 6 \\
+\ 8_1\ 7_1 9 \\
\hline
1\ 0\ 2\ 5
\end{array}
$$

We learnt it mechanically so that it became obvious, but it is far from obvious to a child. For example, in the sum above, adults say: '6 add 9 is 15, put down the 5, carry the 1. Now 4 and 7 and 1 is 12, put down the 2, carry the 1 . . .' and so on. But '4 and 7 and 1 is 12' actually means '40 add 70 add 10 is 120' and the next column is not '1 add 8 add 1' but '100 add 800 add 100'. We understand this code, but without a solid foundation, children don't.

Vertical subtraction – the parents' method

If vertical addition can confuse a child, then vertical subtraction (the way older generations were taught) can be an absolute minefield.

Here is one method you might have been taught for subtraction, using the example 784 − 356 =

$$
\begin{array}{r}
7\;\,8\,{\scriptstyle 1}4 \\
-\,3\,{\scriptstyle 1}5\;\,6 \\
\hline
4\;\,2\;\,8
\end{array}
$$

Starting on the right you'd say, '6 from 4 won't go, borrow one, 6 from 14 is 8, pay back the 1, 5 + 1 is 6, 6 from 8 is 2 . . .', and so on. (Actually, depending on what age you are and what school you went to, there are subtle differences about where you put the small '1' that is paid back – as described in Tom Lehrer's 1960s song 'New Math').

Younger mums and dads might have been taught to 'adjust' the top number:

$$
\begin{array}{r}
7\;\,{}^{7}\!\!\not{8}\;\,{}^{1}4 \\
-\,3\;\,5\;\,6 \\
\hline
4\;\,2\;\,8
\end{array}
$$

Starting on the right you'd say, '6 from 4 won't go, take ten from the 8 [which is actually 80] and add it to the 4 to make 14. Six from 14 is 8, 5 from 7 is 2 . . .' With echoes of the older method above, some people were taught to 'borrow' 10 from the 8, but nothing is actually being borrowed in this method, it's just that the 784 is split up into 770 + 14.

Confusion with vertical calculations

Young children starting to record maths don't see the need to put symbols together in a particular order. So $3 + 4$, $4 + 3$, or even $3\ 4\ +$ or $+\ 3\ 4$ all mean the same to the youngster: add together 3 and 4. Similarly, they have no problem with $7 - 3$ or $3 - 7$: both simply mean subtract 3 from 7 (and if we say take 3 from 7, then $3 - 7$ is quite a sensible move into recording). Then a helpful adult (let's call her Miss Prim) comes along and says, 'You can't take a bigger number from a smaller number, so don't write $3 - 7$ write $7 - 3$.' But strictly speaking she's wrong, you *can* take a bigger number from a smaller — no doubt it sometimes happens on your bank statement. Mathematicians invented negative numbers precisely so that they could give answers to calculations like $3 - 7$. It's just that the child cannot do it *yet*.

Later on, another helpful adult (call him Mr Proper) shows them how to do column subtraction, with a calculation like:

$$
\begin{array}{r}
468 \\
- 245 \\
\hline
223
\end{array}
$$

'Start with the units,' Mr Proper's instruction goes. '8 take away 5 is 3, 6 take away 4 is 2, 4 take away 2 is 2.' (Actually we are taking 40 away from 60 and 200 from 400 but that's another story.)

Later, the child comes across:

$$452$$
$$- 289$$

and the voices of Prim and Proper begin to merge:

Proper: Start with the units.
Child: 2 take away 9.
Prim: You can't take a bigger number from a smaller one.
Child;   If I can't do 2 take away 9, then I can do 9 take away 2, that's seven.

And so, following the advice they have been given, the child starts to write down things like:

$$452$$
$$- 289$$
$$237$$

These rules can be particularly confusing for children, which is why this way of working is held back until they really understand what they are doing. The quizzes opposite give examples of children getting the answer wrong using the 'old' method and of children getting it right using their own methods.

Inside kids' heads:
how did they get these *wrong* answers?

Take a look at these subtraction calculations done by children. Although the answers are wrong, there are sensible reasons as to why the children got them wrong. Can you figure out what the child's thinking might have been?

A
$$
\begin{array}{r}
543 \\
-287 \\
\hline
344
\end{array}
$$

B
$$
\begin{array}{r}
201 \\
-\ 97 \\
\hline
14
\end{array}
$$

A The child probably has said to herself, '3 take away 7 you can't do, so 7 take away 3 is 4' and so on.

B This is a tricky one to figure out. The child realised that to take 7 away from 1 in the right-hand column she had to 'borrow' 10, but since there is a zero in the tens column of 201, she moved to the left and borrowed from the 2 in the hundreds column instead. (Another possibility is that she had been practising lots of subtractions of two-digit numbers and thought, incorrectly but sensibly, that the 'borrowing' always has to be done from the digit on the far left.)

The children making mistakes like these haven't developed a strong number sense. They will have little understanding of what is going on with the old method of crossing out and 'borrowing'. The number line and partitioning methods that we looked at in the last chapter make more sense to them.

(We heard a wonderful story about a child who took borrowing to an extreme. In carrying out a sum like

543 − 287, the child realised that taking 7 from 3 required some borrowing, but instead of borrowing from the next column, he decided to borrow 4 from the date – written in numbers – at the top of the page! He ended up with the right answer, as long as the teacher was prepared to accept that today's date had mysteriously reduced by 4.)

Addition by partitioning

Late in primary school, your child does begin to learn to do paper-and-pencil 'vertical' calculations, but they may initially look a bit different from what you are used to. Many schools now introduce vertical methods by first of all setting them out in an 'expanded form' that makes the method clearer and that can be gone back to if forgotten. Here's how addition might be set out:

You would write:	*Child might be taught:*
452	400 50 2
+ 289	+ 200 80 9
	600 + 130 + 11 = 741

Test yourself

i) *Addition using 'partitioning'*

a. Calculate 147 + 242 using partitioning.

b. Calculate 368 + 772 using partitioning.

Subtraction by partitioning

Subtractions can be set out similarly and then rearranged to make the subtraction in each column possible.

$$
\begin{array}{r}
452 \\
-\,289 \\
\hline
\end{array}
\qquad
\begin{array}{rrr}
400 & 50 & 2 \\
-\,200 & 80 & 9 \\
\hline
\end{array}
$$

At this stage, there are still challenges in subtracting 9 from 2 and 80 from 50, so the numbers in the top row can be reorganised (**partitioned** differently) so that each subtraction can be carried out.

Is the same as . . .   Or . . .

$$
\begin{array}{r}
452 \\
-\,289 \\
\hline
\end{array}
\quad
\begin{array}{rrr}
400 & 50 & 2 \\
-200 & 80 & 9 \\
\hline
\end{array}
\qquad
\begin{array}{rrrl}
300 & 140 & 12 & \text{(still 452)} \\
-200 & 80 & 9 & \\
\hline
100 & 60 & 3 & = 163
\end{array}
$$

It's now a short step to the traditional compact form.

Test yourself

ii) *Subtraction using 'partitioning'*

a. Calculate $847 - 623$ using partitioning.
b. Calculate $721 - 184$. (Instead of doing it as $700 - 100$, $20 - 80$ and $1 - 4$, find a way of doing it so that you don't have to deal with any negative numbers.)

Inside kids' heads:
how did they get these *right* answers?

Children are using a much wider variety of approaches for doing subtractions than older generations. The good news is that this is much more empowering for the children. The bad news (for parents) is that it requires more effort to work out what the children are doing. Before you do this subtraction quiz, have a go at the sum 56 − 38 using a pencil and paper.

Now look at the same sum tackled by three different children. All the children got the right answer, but each child used a different method. Can you work out the child's method?

```
A   56        B    56       C    56
   -38            -38           -38
    26            ─2             2
    20            20            16
    18            18            18
```

A The child says to herself something like: 'From 56 take away 38. Subtract 30 from 56, that's 26. I've still got another 8 to subtract. Subtract 6 from 26, that's 20. Two more to subtract from the 20, that's 18.' Because the girl didn't write down one of her steps, you may have found this method hard to follow. It's fine if these are simply jottings that she is using to work out an answer, but if she wants somebody else to check her working, she should be encouraged to write out her workings like this (for example):

$$56 - 30 = 26$$
$$26 - 6 = 20$$
$$20 - 2 = 18$$

B The child has used his understanding of negative numbers to take the 8 away from thc 6 – that's negative two – then the 30 from the 50 – that's 20. Twenty add negative 2 is 18. (Wow!)

C The child has actually done more addition than subtraction, by figuring out what needs to be added to 38 to make 56: '38 add 2 is 40, 40 add 16 makes 56, 2 add 16 is 18.'

Don't be surprised if you find these methods confusing – the point is that such methods make sense to the child who invents it. And remember, the children did all get the right answer!

The methods used in this quiz were all invented by children who have developed what mathematicians like to call 'number sense' – they have a 'feel' for how subtraction works, they are happy to 'play' with the numbers involved and are not simply following someone else's rules.

Is the answer right?

With any calculation, be that addition, subtraction or some-thing more complicated, there is a final step that children need to learn – checking that the answer makes sense. It takes time to develop this skill, since most children having just completed a sum regard it as finished, and are desper-ate to get on to the next task.

Checking a calculation doesn't necessarily mean going through all of the workings again. Some of the best checks are ones that establish there must have been a mistake, even if you don't know what exactly the mistake was. For example, the answer to $27 + 42$ cannot be 843 because both numbers are smaller than 100. We don't know what the mistake was, but we know for sure that there was one! Try to help your child to look out for obvious errors by developing a habit of asking: 'Does this answer look sensible?'

Test yourself

iii) *Why must these answers be wrong?*

Without doing the whole calculation, how do you know these answers must be wrong?

a. $3,865 + 2,897 = 6,761$
b. $4,705 + 3,797 = 9,502$
c. $3,798 - 2,897 = 1,091$

SIMPLE MULTIPLICATION AND TABLES

Q. One of these numbers below is a multiple of 5.

Put a ring around it.

17   8   52   35   22

On closer examination you can see that the child had
drawn a (diamond) ring around the word 'it'.

From addition, it is just a short step to multiplication – but it is now that problems multiply too – for children, and also for their parents. For children, the main problem is grappling with ideas that are increasingly abstract, while for many parents it is now that unfamiliar methods and new language really begin to rear their heads.

Multiplication falls naturally into two parts: committing the basic multiplication facts to heart (traditionally known as tables) and learning tools to do multiplication of larger numbers. This chapter is dedicated to the tables.

Newspapers regularly trot out scare headlines claiming that schools are no longer teaching your kids the times tables. Well, it's not true. We've never come across a school that doesn't want children to know their tables, or *multiplication bonds* as some educationalists now like to call them.

Where the popular press and schools sometimes part company is on the best means of getting children to commit all this to memory.

Common problems children have with basic multiplication and tables

1. Not knowing the answer to 7×8.
2. Not recognising that a problem is about multiplication (because the question doesn't explicitly say, 'What is 8 multiplied by 4?')
3. Not realising that if you know $4 \times 9 = 36$, then you also know the answers to 9×4, $36 \div 4$ and $36 \div 9$.
4. Not having any methods to fall back on to figure out the answer if they forget a bit of a table.

5. Always counting 4, 8, 12, 16, 20, 24 rather than committing $4 \times 6 = 24$ to memory.

Early multiplication and learning tables

Children first encounter multiplication very young, as a shortcut to adding up. Instead of adding seven plus seven plus seven plus seven, it's very handy to be able to remember that 4 lots of 7 is 28. So important are these basic calculations throughout maths – and indeed most other subjects – that it is vital to be able to commit them to memory.

Here, in table form, are the 100 multiplication facts from 1×1 to 10×10 that children need to be able to quickly recall.

1	2	3	4	5	6	7	8	9	10
2	4	6	8	10	12	14	16	18	20
3	6	9	12	15	18	21	24	27	30
4	8	12	16	20	24	28	32	36	40
5	10	15	20	25	30	35	40	45	50
6	12	18	24	30	36	42	48	54	60
7	14	21	28	35	42	49	56	63	70
8	16	24	32	40	48	56	64	72	80
9	18	27	36	45	54	63	72	81	90
10	20	30	40	50	60	70	80	90	100

And that's it. Learn that lot and you have a great foundation for doing arithmetic – and mathematics – for the rest of your life.

If only it were that simple! Learning times tables is a source of anguish for parents – particularly the frustration that children don't seem to commit them to memory like they used to.

The language of multiplication

Before launching into the world of multiplication with your child, it's worth stepping back and realising that a simple multiplication calculation can be described in a surprising number of ways. Take 3 × 4, for example. You can say this as:

- *three* fours (or *four* threes)
- three *times* four
- three *multiplied by* four
- *the product of* three and four
- three *groups of* four
- four *lots of* three

Slowly, your child will begin to realise that these all mean the same thing, and that they all mean multiplication, but that's far from obvious at the start. You can help by casually using different language when talking about a multiplication, instead of repeating yourself. For example: 'So what's three fours; what's three times four?'

(Just for completeness, when you move into the world of algebra, where numbers are represented by letters, there are a couple more ways in which multiplication is represented. Because the letter 'X' looks like a multiplication sign, you might see multiplication represented by a dot, or more often using no symbol at all! For example '4(b − 3)' means 4 multiplied by (b − 3). But your child probably won't encounter this sort of terminology until their teens, so unless your child's maths is advanced, don't confuse matters by introducing this idea earlier.)

Learning by chanting

Most adults learnt tables by chanting. 'One four is four, two fours are eight, three fours are twelve . . .', and so on. Typically, the tables were learnt in order – first you learned your twos, next your threes . . . all the way up to your twelves. There were two ways of learning your tables – the number of the table could go first or second. So instead of learning your fours as 'one four is four, two fours are eight . . .' you might have learnt them as 'four ones are four, four twos are eight, four threes are twelve . . .'

Regardless of which way round you did it, chanting can certainly work. Perhaps the sing-song sound of 'four eights are thirty-two' still echoes in your mind as you do the sum, and children are certainly good at learning sounds by heart. The other advantage of chanting is that it is repetitive, and there is no doubt that if you repeat something often enough then it just sticks. So if your child likes chanting tables, then do encourage them in this.

However, not all children like chanting. And there's another snag with chanting, too. Since the chants of the times tables don't rhyme, it would be possible to chant a times table with all the right rhythm yet get all the answers wrong. 'Four eights are thirty-six' trips off the tongue just as easily as 'four eights are thirty-two'. There's no immediate way of knowing which is the right answer, apart from faith in your own memory. Nowhere does this problem with faulty recall show up more than in the calculation: 7×8.

If you want to catch someone out on whether they really know their tables, seven eights is the classic question to ask. It's the one that most people find hardest to remember. Seven eights is 56. Or is it 54? Or 58?

QUICK TIP
To remember 7×8, just remember 5, 6, 7, 8: $56 = 7 \times 8$

The order for learning tables

Learning tables in order – first the twos, then threes, then fours and so on – is not the most effective way to learn them. The most natural order for children to learn their tables is to start with the easiest and work up to the hardest. This order makes more sense:

● Tens (10, 20, 30 . . .) which children learn as a natural part of counting.
● Fives, because of fingers and toes.
● Twos. Pairs, even numbers and doubling are familiar ideas to young children.
● Fours (which are just double the twos) and eights (double the fours).
● Nines (there are nice shortcuts, see below).
● Threes and sixes.
● Sevens.

You'll find tips to help learn these tables later in the chapter.

Why 3 x 7 equals 7 x 3

A big idea in helping children to learn their tables is learning that the order of the numbers doesn't matter: 3×7 has the same answer as 7×3. Mathematicians love this big idea so much that they gave it a fancy name: the *commutative* law (which has the same origins as 'commuter' – both based on the notion of back-and-forth).

For adults this big idea that multiplication is commutative usually seems obvious (though see the VAT question on page 196). Not so for children. It takes a while for the idea to become clear, as their early experiences of multiplication don't usually make this explicit. If Jo has three packets of sweets with seven sweets in each packet and Sam has seven packets of sweets with three sweets in each packet, it's not immediately obvious that Jo and Sam both have the same number of sweets. (And offered the choice between these two, young children will often opt for the seven packets in the expectation of getting more).

One of the best ways of demonstrating to a child why 3×7 equals 7×3 is by using an *array*. Array is a word that may not have cropped up in your maths education, but these days it features prominently in the vocabulary of school maths. It's the formal mathematical word for a set of numbers or shapes laid out in a rectangle. Here for example is an array of 3 rows and 7 columns:

Arrays are an extremely important idea, because they are a simple, visual aid for helping children to understand how multiplication and fractions work. How many dots are there in the 3 by 7 array? Three rows of 7 make 21. In other words, arrays are a simple way of representing a multiplication, in this case $3 \times 7 = 21$.

Now what if you draw the array in two different ways?

The first array shows 3×7, the second 7×3. (It is conventional to 'read' these diagrams as rows then columns. Some people remember this by saying 'The Pope is **R C**'). It's obvious that there must be the same number in each array without counting all the dots, since rotating the first array through a quarter turn makes it look the same as the second one. In other words $3 \times 7 = 7 \times 3$.

And indeed whatever your array (or whatever the multiplication sum), the answer is the same whichever direction you look at it. 247×196 is the same as 196×247, and you only have to think of the array idea to convince yourself of this.

QUICK TIP

Look out for arrays around the house and in the street. Point them out to your child and talk about them. Look at the plastic tray holding biscuits in the tin. It's 3 by 4. What if we turn it round? Now it's 4 by 3. See the windows on the side of that building. Why is it a 5 by 4 array? Can it also be a 4×5 array? In fact, once you start looking for arrays you'll discover that they are just about everywhere.

Halving your tables

Once you have accepted the idea that 3×7 is the same as 7×3, it greatly reduces the number of multiplication facts you need to remember. If you commit 3×7 to memory, then you have the answer to 7×3 as a bonus. It's the mathematical equivalent of BOGOF (buy-one-get-one-free), which becomes ROGOF (remember one, get one free). The knowledge of this turnaround rule reduces the 100 multiplication facts to 55 (not exactly half because of the square numbers such as 3×3 and 7×7 which don't have partners).

You can see how this saving takes place if you take another look at the 100 numbers in the ten times tables on the following page:

1	2	3	4	5	6	7	8	9	10
2	4	6	8	10	12	14	16	18	20
3	6	9	12	15	18	21	24	27	30
4	8	12	16	20	24	28	32	36	40
5	10	15	20	25	30	35	⑩	45	50
6	12	18	24	30	36	42	48	54	60
7	14	21	28	35	42	49	56	63	70
8	16	24	32	④	48	56	64	72	80
9	18	27	36	45	54	63	72	81	90
10	20	30	40	50	60	70	80	90	100

Each of the numbers that appears above the diagonal dotted line (for example 5 × 8 = 40) also appears below the dotted line (8 × 5 = 40). The dotted line is a line of symmetry. (Notice anything about the numbers on the dotted line? See page opposite.)

1	2	3	4	5	6	7	8	9	10
	4	6	8	10	12	14	16	18	20
		9	12	15	18	21	24	27	30
			16	20	24	28	32	36	40
				25	30	35	40	45	50
					36	42	48	54	60
						49	56	63	70
							64	72	80
								81	90
									100

QUICK TIP

Children usually start learning their tables by using counting patterns. To figure out 8×4 they will go 4, 8, 12, 16, 20, 24, 28, 32. If you know that 8 times 4 is the same as 4 times 8 then 8, 16, 24, 32 is quicker. In Japan children are explicitly taught to 'put the smaller number first'. 7 lots of 3? Don't do that, do 3 lots of 7.

Learning the square numbers

Numbers multiplied by themselves (1×1, 2×2, 3×3 and so on) are known as the square numbers. That's because they are the numbers that fill a square array. If you look back at the diagonal of the times table grid opposite you'll see that these are the square numbers.

The square numbers crop up so much in maths later in school that they are worth committing to memory separately from the times tables.

Square numbers have an interesting pattern that you can explore with your child. As you progress through the square numbers . . .

$$1, 4, 9, 16, 25, 36, 49, 64, 81, 100 \ldots$$

. . . look at how much they increase each time:

Square numbers:	0	1	4	9	16	25	36	49 ...
Differences		1	3	5	7	9	11	13

This curious connection between the square numbers and the odd numbers is a nice example of how different number patterns connect to each other in maths.

Tens and fives

The first and easiest table to learn is the tens – 10, 20, 30, 40 . . . is an easy extension of 1, 2, 3, 4.

Children also find the 5 times table relatively easy, helped by having hands and feet that represent four lots of five. It also helps that numbers in the 5 times table always end in 5 and 0, a simple pattern that enables a child to instantly recognise if a number is in the table. (And this applies to big numbers too: we know 3,451,254,947,815 is in the five times table even though it is too long to fit onto a calculator display to check it out.)

Learning by doubling – the twos, fours (and eights)

Children find doubling easy. It's probably something to do with two hands and the five fingers on each. Lend them your hands and all the doubles up to double 10 are quickly learnt.

But they don't always connect doubling with multiplying by two. A child may know that double 6 is 12 but asked what 6 times 2 is and she has to count up 2, 4, 6, 8, 10, 12. The thing to do here is to remind them that 6 times 2 has the same answer as 2 times 6 and talk about 2 times 6 being the same as double 6.

So if your child gets good at doubling then they know their 2 times table. What they won't immediately recognise is that they then also can quickly figure out the 4 times table – just double and double again.

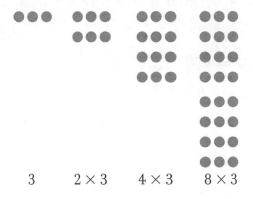

6	2 × 6	4 × 6

This idea extends to the eights. To find 8 lots of 3, for example, you need to double (to make 6), double again (12) and double again (24).

3	2 × 3	4 × 3	8 × 3

The great thing about these approaches is that your ability doesn't run out when the table runs out.

Doubling 4 times will give you 16 times a number, 5 doubles gives you 32 times a number, and so on. What is 32 times 18? We just need to double 18 five times. 18, 36 (× 2), 72 (× 4), 144 (×8), 288 (× 16), 576 (× 32).

Game: Double Snakes & Ladders

You can adapt any game that involves dice by changing the rule so that what you roll counts as double. This has several advantages: the children like the idea of being able to travel twice as far with a throw of the dice; they become familiar with their 2 times table; and (for parents keen to get on with other chores) the game is over twice as quickly.

Test yourself

i) Working out 8 x 7

Can you work out the sum 8×7 using the doubling method?

9 times table

One way of getting to grips with the 9 times table is to do ten times and take off the excess. What is 9 times 7? Ten times 7 is 70, take off 7 is 63. A quick sketch of an array can help fix this idea in mind.

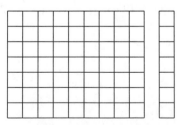

$$7 \times 9 = (7 \times 10) - 7 = 63$$

If you've only learnt the 9 times table up to 9 times 10, then 9 times 25 will stump you. But 10 times 25 is 250, take off 25, 225. Nine times 25 is 225.

Test yourself

ii) Compensation method

Can you work out 9×78 in your head, using the compensation method (multiplying by 10 and then taking 78 off)?

The 9 times table finger method

There's another neat way of figuring out the 9 times table. It uses your fingers and children love it. Hold out your hands face up and imagine the digits (fingers and thumbs) are numbered from 1 to 10, 1 being the little finger on your far left, 10 the little finger on your far right.

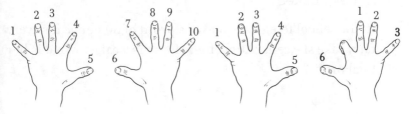

To multiply by 9, bend forward the digit with the number you want to multiply. Nine times 7, say? Bend over the finger you've mentally numbered 7.

Look at your hands: the number of fingers to the left of the bent finger gives the number of tens in the answer; 60 in this case. The number of fingers to the right gives the units: 3. $9 \times 7 = 63$. Try it, it; it works for all the numbers.

The 3 and 6 times table

For children, the 3 times table is actually one of the hardest to learn. There are no real shortcut tips for learning multiplying by 3 (some people suggest doubling the number and then adding it on, so 3 × 7 is double 7, that's 14 add on 7 that's 21, but it's really no quicker than going 7, 14, 21). There's no getting round the fact that the 3 times table just has to be memorised. So for this reason it's better to do the others first so that your child has built up their confidence.

The 6 times table follows straight on from threes: once again, it's a case of doubling. If you can multiply by 3, simply double the answer to multiply by 6. So 3 × 7 is 21, 6 times 7 is 42.

Test yourself

iii) Fairy cakes

At the school fete, Mary makes a profit of 60 pence for every fairy cake she sells. Altogether she sells 9 cakes. What is her total profit?

The 7 times table – a dice game

All that is left is the 7 times table. Here's the good news. If your child has got to grips with all the other tables described above, then you don't need to learn the seven times at all – everything has been covered in the other tables.

However, for completeness' sake your child may want to learn the 7 times table, so here's a game that can speed that up. You need as many dice as you can find – ten is a good number. You can perform this as a challenge. Say that you are going to race your child to see who can add up the numbers on the dice faster. However, to give them a chance, you are going to allow them to choose how many dice they want to roll. And to give them even more of a chance, they only have to add up the numbers that are on top of the dice, while you have to add the numbers on the top AND the bottom of the dice.

Get them to select at least two dice, and put them in a shaker (a mug makes a good shaker). All you need to know is how many dice have been chosen.

As soon as they roll the dice, you can work out immediately what the total of the numbers on the tops and bottoms of the dice will be! How? Simply by multiplying the number of dice by seven. So if there are three dice, the top and bottom numbers on the three dice will add to 21. (The explanation, of course, is that the opposite sides of a dice always add to 7.)

They will be so amazed at your speed of calculation that they'll want to learn this method for themselves.

Test yourself

iv) Match up the sums

Match up the calculations on the left with the equivalent calculations on the right:

5 × 4	8 × 6
7 × 12	14 × 3
6 × 7	3 × 18
3 × 16	2 × 10
6 × 9	4 × 21

Elevens and twelves

'Wait a minute,' some parents cry, 'I learnt my tables up to 12 × 12. Why is my child let off more lightly?' In the days of imperial measurements and pre-decimal currency, people needed to work up to 12 × 12 when there were 12 pennies in a shilling and 12 inches in a foot. This is ancient history for the kids of today, who, brought up in the metric system, only need to deal with multiplying up to 10.

Except . . . twelves do crop up – a lot of people still work in inches (in America it's the standard) and eggs are still sold by the half-dozen and dozen.

And there's more to it than that. Once children begin to get comfortable multiplying numbers that are larger than ten, they start to get a feel for big multiplication sums. Knowing your 11 and 12 times tables can introduce intriguing patterns that might be missed if you stop at ten. Here is the grid for all the tables up to 12.

1	2	3	4	5	6	7	8	9	10	11	12
2	4	6	8	10	12	14	16	18	20	22	24
3	6	9	12	15	18	21	24	27	30	33	36
4	8	12	16	20	24	28	32	36	40	44	48
5	10	15	20	25	30	35	40	45	50	55	60
6	12	18	24	30	36	42	48	54	60	66	72
7	14	21	28	35	42	49	56	63	70	77	84
8	16	24	32	40	48	56	64	72	80	88	96
9	18	27	36	45	54	63	72	81	90	99	108
10	20	30	40	50	60	70	80	90	100	110	120
11	22	33	44	55	66	77	88	99	110	121	132
12	24	36	48	60	72	84	96	108	120	132	144

Notice that the number 8, for example, appears four times in the grid, while 36 appears five times. If you join up the '8 squares' to each other they form a smooth curve. The same for the 36 squares. Indeed, if any number appears more than twice, you can join them up in a curve of similar shape – and if you draw all of these curves, they never cross over each other. (The shape of the curve is known as a hyperbola.)

You can set your child off on an investigation that might – just might – keep them busy for half an hour or more. Print off a few copies of the 12 × 12 grid, then ask them to do the following:

● Colour in all the even numbers red and the odd numbers blue
● Find which numbers appear the most often
● How many different numbers appear in the grid?
● Which is the smallest number that doesn't appear in the grid? Which other numbers between 1 and 100 don't appear in the grid?

The elevens trick

The eleven times table has the simplest pattern of all.

$$1 \times 11 = 11$$
$$2 \times 11 = 22$$
$$3 \times 11 = 33$$
$$4 \times 11 = 44$$
$$5 \times 11 = 55$$
$$6 \times 11 = 66$$
$$7 \times 11 = 77$$
$$8 \times 11 = 88$$
$$9 \times 11 = 99$$

But what happens for bigger numbers? There's an appealing little trick that makes it easy to multiply any number between 10 and 99 by eleven.

● Think of any number between 10 and 99 – let's say 26, for example.
● Now split it in two so there is a gap in the middle. 2 _ 6
● Add up the two digits of your number. $2 + 6 = 8$ and put that number in the middle, 2 8 6.

And that is the answer! $26 \times 11 = 286$.

Be careful, though. What happens if you multiply 75×11?

● Split the number 7 _ 5
● Add the digits: $7 + 5 = 12$
● Put 12 in the middle you get the answer 7,125, which is clearly wrong!

What's wrong? There is a catch if the digits of your original number add up to 10 or more ($7 + 5 = 12$). In this case, you have to add the '1' to the first digit. So 75×11 is not 7,125 but $(7+1)25$, or 825. So it's not *quite* such a neat trick after all.

Test yourself

v) Elevens

Work out in your head:

a. 33×11
b. 11×62
c. 47×11

Game: Beat the calculator

The aim of this game is to get quick at tables. You need: a pack of playing cards with the picture cards removed and a calculator.

- Decide who is going to be the first person to use the calculator.
- Shuffle the cards and turn over the top two.
- The player with the calculator has to multiply together the two numbers on the cards and **must** use the calculator, even if they know the answer (yes, this can be very frustrating).
- The other player has to multiply the two numbers mentally.
- A point to the player who gets the answer first.
- Swap over after ten calculations.

MULTIPLICATION
BEYOND TABLES

Q. Sam has £1 in his pocket and apples cost
30 pence each. How many apples can Sam
buy?
Show how you got your answer.

A. 3.
Nickie told me

Once children have got the times tables under their belts, they start to move on to learning methods and strategies for multiplying and dividing larger numbers. As we saw in the chapter on addition and subtraction, the paper-and-pencil methods taught today are rather different from those of a few years ago. While children make errors with addition and subtraction paper-and-pencil methods, they are even more prone to errors with long multiplication and division. One reason for this is that they come to long multiplication having done a lot of adding and subtracting large numbers. Because the calculations are set out in a similar fashion, children often apply rules that they learnt for addition to multiplications: sadly these do not always give the right answer. So go slowly with multiplication and division – your child may think they know what is going on, when they really don't.

Common problems children have with multiplication beyond tables

1. Making mistakes using techniques that they learnt mechanically, without understanding what they were doing.
2. Thinking that multiplication means multiple adding (when often it is about ratios).
3. Assuming that multiplying always makes things bigger (so they are stumped when they discover that multiplying by $\frac{1}{2}$ makes things smaller).

Bigger multiplications – why the methods have changed

Once children are confident with their tables, they are ready to start doing bigger multiplications. In times past, this meant immediately launching into what is usually known as *long multiplication*.

There's a popular myth that there was a time when almost all school children could successfully do long multiplication, getting it right 99 if not 100 per cent of the time. But it is just that – a myth. While some children flourished with this compact, centuries-old technique, many others really struggled. Even if children could use the method to come up with the right answer, they often had little idea of why the method worked, and were doing little more than operating a black box to see what would come out at the far end. And once they were out of practice, they would forget important parts of the technique and errors would creep in. Others never got it in the first place.

Long multiplication is still taught, but usually only after children have been through several other stages first, building their understanding as they go.

Long multiplication – the parents' method

Here is a reminder of long multiplication, for the calculation 36 × 24:

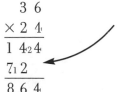

$$\begin{array}{r} 3\ 6 \\ \times\ 2\ 4 \\ \hline 1\ 4_2 4 \\ 7_1 2 \\ \hline 8\ 6\ 4 \end{array}$$

Some people (including us) were taught to leave this space blank. Others were taught to put a zero here so that it reads 720. Putting in the zero is a good idea because it reduces the chance of an error later.

The procedure is to multiply by the digits of 24 in turn, working from the right. The 'script' that you follow as you do the calculation probably sounds like this: 'Four sixes are 24, 4, carry 2, four threes are 12 plus 2 is 14 . . . 144. Move one column to the left, two sixes are 12, carry 1, two threes are 6 plus one is 7, 72.' You then add 144 + 72 (which is really 720, but in our example the zero hasn't been recorded) to give the answer.

Which is great, if you follow the rules correctly. Unfortunately, not all children do.

Inside kids' heads: how did they get these *wrong* answers?

A
$$\begin{array}{r} 3{,}6 \\ \times\ 3 \\ \hline 128 \end{array}$$

B
$$\begin{array}{r} 36 \\ \times\ 24 \\ \hline 24 \\ 600 \\ \hline 624 \end{array}$$

C
$$\begin{array}{r} 36 \\ \times 24 \\ \hline 24 \\ 72 \\ \hline 96 \end{array}$$

A The child correctly multiplied 3×6 and carried 1. She remembered that this 1 has to be added at some point, but unfortunately she adds it at the wrong stage: her thinking goes, '3 plus 1 makes 4, then multiply 3×4 to get 12.'

B The child is treating the long multiplication as if it is column addition. In column addition you add the units and then add the tens which gives you a correct answer to $36 + 24$. His logic is that this could also work for multiplication – multiply the units ($4 \times 6 = 24$), multiply the tens ($20 \times 30 = 600$), and add the answers.

C The child has not appreciated that they are multiplying by 20 rather than by 2. Unfortunately the way we talk about long multiplication doesn't always help here. It's quicker to say, 4 times 6 is 24, 4 times 3 is 12, put down a zero, 2 times 6 is 12, 2 times 3 is six.

It's because of common errors like these that schools today take a different route to multiplication.

Step 1: Multiplying using arrays

The first step beyond tables is to learn how to multiply two-digit numbers by single-digit numbers, for example 3×14.

The most basic method is based on arrays, which we explained on pages 125–6. The calculation 3×14, can be represented as an array of dots like this:

A child can simply count the dots if he wants to, but if he knows his tables, the array can be broken down into parts that a child can calculate easily, by partitioning (that word again) 14 into 10 + 4, like this:

This makes it obvious that 3×14 is the same as 3×10 plus 3×4, or 30 + 12.

Step 2: Drawing boxes

Instead of showing all the dots (which might encourage children to count them) children are next encouraged to represent the array of dots as boxes, with the number of dots indicated along the top and down the side.

3 × 14 becomes:

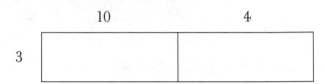

Notice how this isn't drawn to scale — it doesn't need to be. We've just created boxes into which the answers can be written, like this:

In other words, 3 × (10 + 4) = 30 + 12 = 42.

Step 3: Drawing a grid

More complicated calculations can be dealt with in exactly the same way, for example 24 × 36 is written like this:

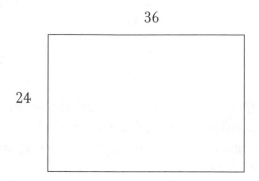

This large box can be broken down into the parts representing the tens and units . . .

... which resembles a grid – and which is why this method is known as the **grid method**. Now to work out 36 × 24, you can add up the number of dots there would be inside each section of the grid.

	10	10	10	6
10	100	100	100	60
10	100	100	100	60
4	40	40	40	24

Add up the 100s, 60s, 40s and 24 and you get the answer 864. That might seem long-winded (because it is!) but (a) it's quicker than counting all the dots, and (b) it's easy to understand exactly what's being done.

Test yourself

i) Grid method 1

Calculate 23×13 by creating a grid and setting the numbers out as tens and units.

Step 4: Working with bigger blocks

Many children will quickly grasp that they can save time by making bigger blocks. A much simpler way to break down 24 × 36 is like this:

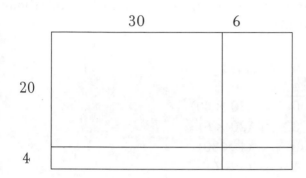

And now there are only four calculations to do:

	30	6
20	20 × 30	20 × 6
4	4 × 30	4 × 6

The answer is found by adding up each of the four calculations in the four parts of the grid: 600 + 120 + 120 + 24.

Step 5: Turning the grid method into 'long multiplication'

Once children become comfortable with the grid method, it is a short step to stop bothering to draw to the grid at all. Instead, the child sets out what was previously a grid as four separate calculations:

$$
\begin{array}{r}
3\ 6 \\
\times \quad 2\ 4 \\
\hline
6\ 0\ 0 \\
1\ 2\ 0 \\
1\ 2\ 0 \\
2\ 4 \\
\hline
8\ 6\ 4
\end{array}
\qquad
\begin{array}{l}
(20 \times 30) \\
(20 \times 6) \\
(4 \times 30) \\
(4 \times 6)
\end{array}
$$

This is very close to the traditional long-multiplication method, and confident children will be shown how to make the final step to the compressed version that was taught to their parents (see page 143).

But why go through all this rigmarole to get there? The reason is that not all children make it all the way to the traditional long-multiplication form. For those children who struggle with multiplication, the grid-method approach ensures that they have a technique that they can understand. And if a child forgets or gets confused at any stage they can 'drop back' to the previous method to fill in the missing pieces. So it's not that the aim isn't to get to the compact form of long multiplication. It's more a case of building up to it in stages so that understanding how it works is as important as being able to do it.

Test yourself

ii) Why must these answers be wrong?

Without doing the whole calculation, how do you know these answers must be wrong?

a. $37 \times 46 = 1,831$
b. $72 \times 31 = 2,072$
c. $847 \times 92 = 102,714$

Bigger multiplications

The grid method doesn't have to stop at multiplying two-digit numbers together. You can use it for any multiplication you like — though needless to say, it gets more cumbersome.

Take the example 134×46. You can set out the calculation like this:

	100	30	4
40	4,000	1,200	160
6	600	180	24

And you can go into the thousands or higher, if you want. Although by the time your child is working out calculations that big, they've almost certainly been shown the traditional long-multiplication method.

Test yourself

iii) Grid method 2

A book costs £9.47 and the school puts in an order for 62 of them. Use the grid method to work out the cost.

How the grid method links to algebra

There's another advantage of the grid method that is quite important, even if it is beyond the level of a primary school child. As a parent, you will remember how in secondary school algebra began to take over, with numbers being replaced by letters. This included expressions such as: $(a + b)$ multiplied by $(c + d)$.

What do you get if you multiply out the brackets? Many parents freeze at this point, until they realise that this is exactly the same procedure as multiplying using the grid method. Think of $(a + b)$ times $(c + d)$ as $(20 + 4)$ times $(30 + 6)$ and you can imagine just setting it out in the grid form and adding up each part:.

	c	d
a	a × c	a × d
b	b × c	b × d

Which is just as it would be with numbers:

	30	6
20	20 × 30	20 × 6
4	4 × 30	4 × 6

So the answer is $ac + ad + bc + bd$. The grid method is a much better foundation for algebra than long multiplication ever was. Just so you know . . .

DIVISION

Q. How many times can you take ten apples away from 35 apples?

A.

$$35 - 10 = 25$$

$$35 - 10 = 25$$

$$35 - 10 = 25$$

$$35 - 10 = 25$$

$$35 - 10 = 25$$

$$35 - 10 = 25$$

D ivision is often the most confusing of all the basic maths functions. The language can be even more baffling than it was for multiplication: '3 into 2 won't go', 'divided by', 'over', 'shared between'. Just how important is long division and why does it cause everyone such problems? And if division is about making things smaller, how come dividing by 0.5 on a calculator gives me a *bigger* answer?

Common problems children have with division

1. Not fully appreciating that division is the inverse of multiplication, so not using the multiplication facts that they know to work out related division facts. For example, if you know that $7 \times 4 = 28$, you also know that $28 \div 7 = 4$ and $28 \div 4 = 7$.
2. Thinking that division is only about 'sharing' ('share 42 apples between 6 people') and not also about repeated subtraction ('put 42 apples into bags with 7 in each bag').
3. Thinking that division always make things smaller: 35 sweets shared between 5 children means each child gets 7 sweets, but there are 5 children so there are still 35 sweets. No sweets have been taken away, they have just been rearranged.

What IS division – sharing or subtracting?

Division is usually introduced as the idea of sharing. Children particularly engage with the idea of sharing sweets (and want to be sure that they get their fair share). So if there is a calculation like this:

$$48 \div 8$$

then it will typically be dressed up in a 'real world' problem as: 'I have 48 toffees that I want to share equally in to 8 bags. How many toffees do I put into each bag?'

But there is another way of interpreting sharing. Compare this problem to the one above: 'I have 48 toffees. I want to put them into bags of 8. How many bags can I fill?'

This can also be solved by calculating $48 \div 8$.

There is a big distinction in the two problems. In the first one, the sharing problem, we know how many toffees there are and how many bags we want to put the toffees into. What we don't know is how many toffees will end up in each bag. To solve this problem you would literally share out 48 objects – set up something to represent the 8 bags and go 'one for you, one for you . . .' until all the toffees were divvied out.

In the second problem the situation is subtly different. There are still 48 toffees but this time you know how many toffees you want to put into each bag, but not the number of bags. To solve this practically you would put out 48 objects and then take away 8 for the first bag, 8 for the next one and so on until all the toffees were used up. This is division as *repeated subtraction* rather than sharing.

Understanding both 'types' of division

It's important that your child is familiar with both types of division problem, the 'sharing' type and the 'subtracting' type. There are two reasons for this.

First, interpreting a question as 'sharing' or 'repeated subtraction' can make a surprising difference to how easy the child finds it to calculate the answer (just as in subtraction, 'take away' and 'finding the difference' changes the way you think about the sum $2001 - 1998$).

Here are a couple of examples that one education expert explored with children:

$$6,000 \div 6 \qquad 6,000 \div 1,000$$

Children who thought of division as sharing find the first problem easy – they can picture in their minds 6 people and imagine giving 1,000 things to each of them. But they find the second calculation difficult, as they cannot cope with trying to imagine 1,000 people. In contrast, children who treated division as repeated subtraction found the second calculation easy – all they had to do was subtract 1,000 from 6,000 as many times as they could, which was 6 times. But repeatedly subtracting 6 from 6,000, wow, that was going a long time. Being flexible in thinking about which version of division to use makes both calculations easy. As does confidently knowing that $1,000 \times 6 = 6,000$ and using the relationship between multiplication and division.

The second reason for understanding both types of division is that when (later in school) children start to divide by fractions, it is really only repeated subtraction that makes any kind of sense.

Dividing by half

What does $16 \div \frac{1}{2}$ mean? Sixteen shared between $\frac{1}{2}$? I can share between two people, but I can't share between half a person!

On the other hand, thinking of it as 'How many times can I subtract $\frac{1}{2}$ from 16?' makes it easy — the answer is 32 times. There are 32 halves in 16. Putting this into a realistic context can help your child's understanding further: a pizzeria sells pizza in half pizza slices. They bake and sell 16 pizzas. How many half slices is that? We deal with this in more detail in the chapter on fractions.

Test yourself

i) Number sequence

The numbers in this sequence decrease by the same amount each time. What are the missing numbers?

43 ☐ ☐ ☐ 7

Prime numbers

A prime number is a number greater than 1 that can't be divided exactly by any other number except 1 or itself without leaving a remainder. Children will often encounter prime numbers for the first time when talking about sharing, which is why we've included them in the division chapter. One way of thinking of prime numbers is to imagine them as the numbers that don't allow you to share your sweets fairly. If you have 15 sweets, then you can divide them exactly between 5 children (3 sweets each) or 3 children (5 sweets each). But if you have 13 sweets, there's no number of children where you can divide the sweets up evenly – except for 1 and 13.

The first few prime numbers are 2 (the only even prime number), 3, 5, 7, 11, 13, 17 and 19. Mathematicians delight in finding prime numbers that are record-breakingly large, and they know that every time they work out the latest 'biggest' prime number, there is certain to be one that is even bigger. (How do they know this? A Greek called Euclid proved it over 2,000 years ago. His proof is beautiful but a bit hard for young children, so we won't include it here.)

Test yourself

ii) Spot the primes

Which of these are prime numbers?

27   37   47   57   67

Factors and multiples

Factors and multiples are not the same thing, though children have a tendency to mix them up. Both are closely connected to division and multiplication, and it's helpful to understand them as they can help you to learn shortcuts in division calculations later on.

Factors are, if you like, the building blocks that multiply together to make a number. Take the number 18, for example. The *factors* of 18 are 1, 2, 3, 6, 9 and 18. These factors can be put into pairs. $1 \times 18 = 18$, $2 \times 9 = 18$, $3 \times 6 = 18$.

Children will often be asked to find all the factors of a number, and it helps to do this methodically by starting at 1 and working upwards. Each time you find a factor you can find its 'buddy'. So for the number 18, 1 is a factor (and its buddy is 18), 2 is a factor (its buddy is 9), 3 is a factor (buddy 6), 4 is not a factor, nor is 5, 6 is a factor and its buddy is 3 . . . but hang on, we've already found 3. There's no point going any higher because we'll just be finding duplicates.

Pairs of numbers may have some factors in common. For example, 18 and 27 both have 1, 3 and 9 as factors. Nine is the *highest common factor* that they share.

Meanwhile, the *multiples* of 18 are 36, 54, 72, and any other whole number multiplied by 18.

Any pair of numbers will have a multiple that they both share. Take 18 and 25. Multiply 18 by 25 and you will get a multiple of 18. Multiply 25 by 18 and you get a multiple of 25. But the answer to these two calculations is the same, and gives you a *common multiple* of both numbers (in this case 450).

All numbers have an infinite number of common multiples, for example 6 and 9 have common multiples of 18, 36, 54 . . . and larger numbers such as 360, 1,800, 90,000 and so on. We can never know what that largest common multiple of two numbers is (that is going into the realm of the infinite) but we can always find the *lowest common multiple* (in the case of 6 and 9 it is 18).

Test yourself

iii) Factor sorting

Write these numbers in the correct places in the diagram:

9 12 15 25 30 90

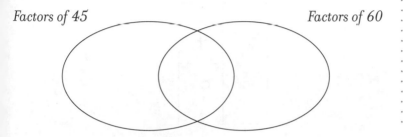

Factors of 45 *Factors of 60*

Division as the opposite of multiplication

How does a child actually *work out* 48 ÷ 8? One way is simply to keep subtracting 8 away from 48 until there is nothing left. And there's nothing wrong with that, except that it is a bit slow. The faster way of doing division is to know your tables. If you want to help your child succeed in division the best thing to do is to make them really confident with multiplication.

How do you know 48 ÷ 8 = 6? Because you are really secure in knowing that 6 × 8 = 48. (We have a theory that no one really ever does a division – instead, they intuitively ask themselves what multiplication gives the answer.)

Game: Division cards

You can help your child here by making a set of cards (cut a sheet of A4 paper into 4) that have a multiplication presented like this:

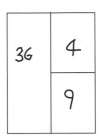

Cover up one of the numbers and work with your child to describe the relationship between the other two numbers in as many ways as you can. In our example, cover up the 4:

- 36 divided by 9 is what?
- 9 times what is 36?
- What must 36 be divided by to get 9?
- How many times can I subtract 9 from 36?
- What is 36 shared amongst nine?

When your child is confident with division based around their tables, they are ready to move on to the big stuff . . .

Divisibility tests

All multiples of 5 end in either a 5 or a zero. All multiples of 2 are even numbers (ending 2, 4, 6, 8 or zero). These patterns can be useful in reverse, in helping to spot whether or not a number will be exactly divisible by another. For example, just by checking the final digit, we know that 872 is not exactly divisible by 5 but it is divisible by 2.

There are three other divisibility texts that can be particularly useful, even though the reason why they work is not entirely obvious:

Divisibility by 3: Add up the digits in the number. If, and only if, the digits add up to a multiple of 3, then the number itself is exactly divisible by three.For example the digits in 211 add to 4, which is not divisible by three. On the other hand, the digits of 174 add to 12 which is divisible by 3, so we know that 174 will divide exactly by 3 (in fact 174 ÷ 3 = 58).

Divisibilty by 6: If the number is even and also passes the test for divisibilty by 3 (see above) then it is divisible by 6. For example 8,412 is divisible by 6 because it is even and the digits add up to 15.

Divisibility by 9: Add up the digits in the number. If, and only if, the digits add up to a multiple of 9, then the number is exactly divisible by 9. So 442 is not divisible by 9 (its digits add up to 8) but 378 is divisible by 9 (its digits add up to 18).

Test yourself

iv) Divisibility tests:

Without doing the whole calculation can you say which of the following divide exactly with no remainder?

a. 28,734 ÷ 2   b. 9,817 ÷ 5   c. 183 ÷ 3
d. 4,837 ÷ 9   e. 28,316 ÷ 6

Long division – the parents' method

Somebody once said: 'Anyone who has done two long divisions in their life has done one too many.'

You might like to think about when you last had to do a long division, other than when helping out with homework! Australia dropped long division from its curriculum many years ago and no one seems to have missed it. In the UK, it is still regarded as the pinnacle of primary maths by some people, so it's likely to be around for a while. Here, as a reminder, is the classic way of doing a long division sum (in this case 517 divided by 24). In this sum 517 is the dividend and 24 is the divisor:

$$
\begin{array}{r}
2\,1 \text{ r } 13 \\
24\overline{)517} \\
48 \\
\hline
37 \\
24 \\
\hline
13
\end{array}
$$

24 doesn't go into 5.
24 does go into 51. 24 into 51 goes 2 times (write 2 on the top line). 2 × 24 = 48 so write 48 under 51.

Subtract 48 from 51 (= 3) and write 3 underneath. Then draw down the next digit, 7, from the dividend to form 37. Divide 24 into 37 – it goes once, put 1 on the top line.

Subtract 24 from 37 (= 13), and since 13 is smaller than 24 and there is nothing left to draw down, 13 is the remainder. Write this on the top line.

We won't go into any more detail about long division here. Why not? If you're confident with long division then the example above should be enough to remind you, and if you're not, it's better to start again from scratch using the methods your children are being taught.

Short division

For calculations where the divisor is small, the full workings aren't essential. So for the calculation 749 divided by 7, you can use 'short division': (incidentally, this method is these days sometimes called the **bus-stop** method because it apparently resembles numbers in a shelter lining up to catch a bus)

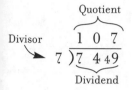

The 'script' that many of us were taught goes something like this:

- 7 into 7 goes once, write down the 1
- 7 into 4 doesn't go, write down the nothing (zero), carry the 4
- 7 into 49 goes 7 times, write down the 7

The answer is 107.

Inside kids' heads: Explain the *wrong* answers

Here is the 'short division' shown above presented as a problem. 'Elinor wanted to cut a piece of ribbon that measured 749cm long into 7 equal pieces. How long would each piece be?'

Here are answers from two children. Can you see why they made mistakes?

A $\dfrac{1\,7}{7\,\overline{)7\,4\,9}}$

$= 1\,7$

B $\dfrac{1\;0\;1}{7\,\overline{)7\,4\,9}}\; r\,2$

A The child has said to herself something like: '7 into 7 goes once, write down the 1. 7 into 4 doesn't go, nothing to write down [whereas she should have written a zero at this point]. 7 into 49 goes 7 times, write down the 7. The answer is 17.'

B The mental 'script' probably went something like: '7 into 7 goes once, write down the 1. 7 into 4 doesn't' go, write down the nothing. 7 into 9 goes once, remainder 2, write down the 1 and the remainder. The answer is 101 r 2.'

The children's incorrect answers are not that far away from the correct script. It's easy to misrecall 'write down the nothing' as 'nothing to write down', or to put a zero above the 4 and move onto the next digit, the 9, and think that the 4 has been finished with (after all the first digit, the 7, was finished with after the one was written down above it).

One of our recurrent messages is that for children not to make mistakes in arithmetic it's important to encourage them to think about numbers rather than digits. As you can see here, 'scripts' that deal with digits are ripe for errors to occur.

Division by 'chunking' (or 'grouping')

Take a look at how a child might typically tackle the calculation 749 ÷ 7 today.

$$
\begin{array}{r}
7 \,\overline{\smash{)}\,749} \\
700 \quad \times 100 \\
\hline
49 \quad \times 7 \\
\hline
107
\end{array}
$$

What is going on here? Just as multiplication can be set out in an expanded form, here the child is using an expanded form for division. They have said to themselves something like:

● How many sevens can I get out of 749?
● Well, 100 sevens is 700 so that's 100 sevens I can get [the child writes × 100 in the right hand column].

- That leaves me with 49.
- I know 7 sevens are 49, so that's another 7 lots of 7 that I can get [she writes × 7 in the right-hand column].
- So that's 107 sevens altogether [adding the 100 + 7 together].

This extended method is sometimes called **chunking** or **grouping**, based on the idea that large 'chunks' or 'groups' are being subtracted from the number being divided into. This approach can also be used for long division.

Test yourself

v) Chunking 1

Divide 336 by 8 using chunking (that is, taking away 'chunks', or multiples, of 8).

. . . .

Inside kids' heads: How did they get these *right* answers?

Children are still being taught long division, but you may not recognise the long division that your child brings home. Their method will almost certainly be based on subtraction, or chunking. The workings for long division that your child is taught may look something like the examples below. Both children have reached the right answer in different ways, but the underlying approach they've used is the same, which is to ask: 'How many times can I subtract 24 from 756?' Can you figure out how they have worked out their answers?

A   24 / 756

        240   10x
        ———
        516
        240   10x
        ———
        276
        240   10x
        ———
        36
        24   1x
        ——
        12

Answer 31 r 12

B   24 / 756

        720   30x
        ——
        36
        24   1x
        ——
        12

31 rem.12

A The child is confident in knowing that 10 lots of 24 is 240 and keeps subtracting 240 (three times) until she gets down to 36. Then she subtracts one more 24, to leave a remainder of 12. Her answer is 10 + 10 + 10 + 1 or 31 (with 12 left over).

B The child mentally went through, '10 twenty-fours is 240, 20 twenty-fours is 480, 30 twenty-fours is 720, 40 twenty-fours is going to be too big, so I'll subtract 30 lots of 240, and then one lot of 24.' While this last solution is a little more efficient than the other one, it's not so much quicker.

In both examples the method is based on what children are confident in doing, rather than them trying to remember a procedure. This is all good training in getting comfortable with handling numbers, and in particular for getting a rough idea of what the answer to a large division sum is. Your child may never have to do a long division manually once they pass the age of 15, but the skills they learn in tackling these problems are a vital part of becoming a numerate adult.

Test yourself

vi) Chunking 2

Divide 739 by 22 using chunking (in other words, remove chunks of 22).

Game: A division mystery

Think of any number between 100 and 999. Enter that number twice on a calculator (for example if you choose 274 then enter 274274 on the calculator). What is the chance that the number you have entered divides exactly by 7? What is the chance it divides exactly by 11? And what is the chance that it divides exactly by 13?

In each case, it might be reasonable to think that it's quite unlikely — after all, only one number in every 7 divides exactly by 7, and only one number in 13 divides by 13. Yet we guarantee that your six-digit calculator number not only divides exactly by 7, but also by 11 AND by 13.

How do we know? Because a number of the form abcabc (such as 274274) is the same as saying abc × 1,001 (in this case $274 \times 1,001$). In other words, abcabc always divides exactly by 1,001. And what numbers divide exactly into 1,001? 7, 11 and 13 – they are its prime factors.

So we can guarantee that whether you chose 872872 or 195195, or any other combination, it was bound to be divisible by 7, 11 and 13. The maths says so.

Test yourself

vii) Why must these answers be wrong?

Without doing the whole calculation, how do you know these answers must be wrong?

a. $223 \div 3 = 71$
b. $71.8 \div 8.1 = 9.12$
c. $161.483 \div 40.32 = 41.36$

BEYOND
ARITHMETIC

FRACTIONS, PERCENTAGES AND DECIMALS

Q. Simplify this $\dfrac{16}{64}$

A.

$$\dfrac{1\cancel{6}}{\cancel{6}4} = \dfrac{1}{4}$$

The child's creative approach has come up with the correct answer, though for the wrong reason. Cancelling out digits like this is not supposed to work!

Parents often name fractions as one of the trickier topics in maths, yet your child will probably be comfortable with the notion of simple fractions from a very early age. By the age of two, children have figured out that birthdays are A GOOD THING, and that knowing how old you are and when your next birthday is coming is a handy bit of knowledge. A child will tell you that they are two and a half, with an intuitive knowledge that this is more than two and less than three, without any instruction in fractions at all.

But many adults will tell you that learning about fractions marks the point at which mathematics began to unravel for them and for their children.

Common problems children have with fractions

1. Thinking that half must always be bigger than a quarter (so why isn't half of £10 more than a quarter of £100?).
2. Thinking that if you chop something into say, five pieces, each piece must be one fifth (even if the pieces are different sizes). To many children, 'half' means 'one of two pieces'.
3. Believing that, say, a quarter of a pie always has to look the same shape.
4. Not realising that 'half', 0.5 and 50% all represent the same fraction.
5. Getting confused over the distinction between '*the* seventh' (as in the seventh in a line) and '*a* seventh' (as in the amount of chocolate you get if you share a bar fairly with six friends).

What is a fraction?

In maths terms, a fraction refers to any number that is not whole that can be expressed as one whole number divided by another. 3 divided by 4 is a fraction, but so also is 10 divided by 3. Fractions written as one number divided by another used to be known as 'vulgar' or 'common' fractions, but these descriptions are rarely used today, partly because the word 'vulgar', which used to be understood as being 'of the people' or 'everyday', now tends to be associated with being rude.

The top part of a fraction is called the numerator and the bottom is the denominator – two terms that people always get confused, and therefore invariably have to be explained again every time they are used ('the numerator, that's the number on the top . . .' and 'the denominator, on the bottom . . .'). One aid to remembering it is with the phrase '**D**enominator is **D**own'.

You may also hear your child talk about two types of (common) fraction:

Proper fraction: A fraction where the numerator (the number on the top) is smaller than the denominator (the number on the bottom), for example $\frac{3}{7}$ or 'three sevenths'.

Improper fraction: A fraction where the numerator is larger than the denominator, for example $\frac{11}{5}$, or 'eleven fifths'. (It's a misleading name, as there is nothing 'improper' about it at all, just as there's nothing vulgar about common fractions.)

Fractions in which the denominator (that's the bottom number, remember) is ten, 100 or another power of ten are known as *decimal fractions* $\frac{1}{10}$, $\frac{3}{100}$ and $\frac{17}{1000}$ are all decimal fractions, but they would usually be written as 0.1, 0.03 and 0.017. One particular decimal fraction carries a familiar everyday name – per cent. All that per cent means is 'divided by 100'. 73 per cent could be written as $\frac{73}{100}$ or 0.73 but is usually written as 73%. There's more about decimals and percentages later in this chapter.

For completeness, your child will also be taught about so-called *mixed numbers*, which is any number expressed as a whole number followed by a fraction. The girl who says her age is $4\frac{1}{2}$ is using a mixed number.

Fraction words

Divide a pizza into two equal parts and you get two halves. Three equal parts are called thirds, four equal parts quarters, and so on. (A bit like our counting system, the naming gets easier as you go on: $\frac{1}{6}$ is a 'sixth' and $\frac{1}{7}$ is a 'seventh'. So why isn't $\frac{1}{2}$ a 'twoth'? Americans do use 'fourths' rather than 'quarters' – except of course the quarter coin, which is a fourth of a dollar!)

Children don't normally have any problem with understanding what to call these parts, but they do often fail to realise the importance of the word 'equal' here. If you break a biscuit in two, a child will often describe the two pieces as halves even though they aren't equal (and they will, of course, be keen to take the bigger 'half'). So you can reinforce with them the correct meaning of fractions by dividing your pizza up and saying 'is that a half?' or 'is that a quarter?' and checking that the pieces are the same size by overlaying them.

For smaller fractions, it is normal to refer to 'eighths', 'tenths', and so on, but beyond twelfths, this becomes a bit of a mouthful. So if you want to say out loud the fraction $\frac{3}{22}$, it's customary to say 'three over 22' or even 'three divided by 22' rather than 'three twenty-seconds'.

Half of what? Use food

Although eventually we come to think about $\frac{1}{2}$ as a number in its own right, when your child is beginning to learn about fractions it is most helpful to always be asking 'a half of what?'.

When grappling with the idea of fractions, and explaining fractions, it's always good to resort to food. There are some who reckon that pizzas were invented purely as a means of introducing fractions, because it is so normal to divide a pizza into fractions such as halves, quarters and sixths. Other foods that can instantly help in tackling fractions are a pack of sausages or a family bar of chocolate (with an array of chunks in rows and columns). Armed with these foods, you are ready to tackle most of the common fraction problems.

Fractions are the answer to fair division sums

One useful introduction to fractions is to think of them as the result of sharing food fairly.

For example:

- Four children share out equally eight sausages. How much sausage do they each eat?

- Four children share out equally three pizzas, how much pizza do they each eat?

Although it seems odd to use the language of 'how much sausage' rather than 'how many sausages' when the answer is 2, the logic to each of these situations is the same: divide a quantity by 4.

$8 \div 4 = 2$ (sausages)
$3 \div 4 = \frac{3}{4}$ (of a pizza)

The connection between these is clearer if we express the answer to the first as a fraction:

$8 \div 4 = \frac{8}{4}$ $(=2)$
$3 \div 4 = \frac{3}{4}$

Seeing fractions as the result of a division calculation helps children enormously. It also makes division the easiest of all calculations:

What is $1{,}234 \div 14$?

The answer is easy: $\frac{1234}{14}$. The answer is in the question!

Inside kids' heads:

Children often get stuck because they think of fractions in a particular way. Can you see why these children made the mistakes they did?

Q1 Three quarters are shaded in the diagram A. How many quarters are shaded in diagram B?

Child's answer: One.

Q2 One quarter is shaded in diagram A. What fraction is shaded in diagram B?

Child's answer: Two quarters.

In the first question, the child thinks of a quarter as being a particular square shape, so since three squares are shaded in diagram A and only one in B, she thinks only one quarter is shaded in B.

In the second question the child has been trapped into thinking of a quarter as being an absolute amount rather than a relative amount. The correct answer of course is that 2 out of 8, or one quarter, of the segments in the second shape have been shaded.

QUICK TIP

To strengthen your child's comfort with fractions, always talk about a fraction of something — half the cake, a quarter of the twelve sweets, a third of a pint — rather than halves, quarters or thirds as abstract ideas. And build on your child's sense of fairness to help them understand that fractional pieces all have to be the same size. The bigger half is OK at tea-time but not OK in the maths class.

Dividing a pizza fairly

Ask two children to share a pizza, and you run the risk that they will argue about which is the biggest piece. The classic solution to this is to ask one child to cut the pizza and the other to choose — both ought now to believe that they have got at least half the pizza. But what happens when there are three children?

The simplest fair solution (well, almost fair) is this: the first child cuts what she thinks is one third and offers it to the second child. The second child accepts if he thinks it is at least one third, otherwise he cuts the rest of the pizza in half. The third child now takes the piece he thinks is the largest. Next the first child takes the piece she first cut, unless it's already been taken, in which case she takes the piece that she thinks is biggest. The second child gets the remaining piece.

Phew . . . so it's not that simple. But at least each child ends up with a piece that they believe to be at least one third of the whole pizza. Actually, not quite. One small snag is that if the first child cuts a piece that is bigger than one third and the second child takes it, then the third child will look on enviously, since he would have taken that piece if he'd had the chance.

Simple pizza fractions in real life can be remarkably complex!

Comparing fractions

Which is bigger, $\frac{5}{8}$ or $\frac{5}{9}$? It's not immediately obvious to a child. But you can make it much more obvious if you think of the fraction as representing sausages being divided between children. The top number is the number of sausages, the bottom is the number of children.

Now a child's intuition comes in useful. If you have five sausages that you are about to divide equally between eight children, and then another child comes along so that you now have to share the sausages between nine children, will the children now get less sausage each or more? Less, of course. So $\frac{5}{9}$ is less than $\frac{5}{8}$. In the same way, if you now cook more sausages, so you have seven sausages between nine children instead of five between nine, the children will now each be getting more. So $\frac{7}{9}$ is bigger than $\frac{5}{9}$.

This sort of sausage reasoning makes it easy to make many fraction comparisons.

Test yourself

i) *Sausage fractions*

Using sausage reasoning, can you determine which fraction in each pair is the larger?

a. $\frac{6}{7}$ and $\frac{5}{7}$
b. $\frac{4}{11}$ and $\frac{3}{12}$
c. $\frac{3}{5}$ and $\frac{4}{7}$

It is only when the numerator and denominator both increase or both decrease that comparing fractions becomes difficult.

Game: Daft domino story

Put a set of dominoes face down on the table. Turn one face up. Decide what fraction the two numbers on the domino will represent. Who can come up with the silliest story that has that fraction in it?

Suppose you turn over ⚂⚄ . That could be three fifths or five thirds (or $1\frac{2}{3}$). Suppose you agree on $\frac{3}{5}$.

'Five hungry monkeys found 3 ripe bananas. Being very fair monkeys they shared them equally. How much banana did they each get to eat?'

'I got five Easter eggs this year. I ate one, then another, then another, then I felt ill. What fraction of my eggs did I eat before getting ill?'

Simplifying fractions

If you want to share a pizza between three people, the easiest thing to do is divide it into thirds:

But that's not how most people divide up pizzas. A third of a pizza is too large and has a tendency to droop when you pick it up. Instead, by instinct, we divide it into six pieces, and give each person two sixths. It is so obvious that two sixths and one third are the same thing, that it seems unnecessary to have to spell this out. Yet this is the basis of the whole idea of reducing a fraction to its simplest terms: $\frac{2}{6} = \frac{1}{3}$. The notion of simplifying fractions crops up all over the place in maths, and the reason for developing the skill of simplifying fractions is that it makes calculations so much simpler.

The secret to simplifying a fraction is to look for numbers that will divide into the top and bottom of the fraction (the highest common factor – see Multiplication and Division). For example, simplify $\frac{10}{15}$:

$\frac{10}{15}$ *The top and bottom both divide by 5. So this is the same as $\frac{2}{3}$.*

Alternatively, you can write out the numerator and denominator in terms of their factors, so:

$$\frac{10}{15} = \frac{2 \times 5}{3 \times 5}$$

Writing a fraction in this form makes it much easier to find the simplest fraction, by cancelling out numbers that appear on both the top and the bottom (in this case, cancel the 5s to make $\frac{2}{3}$).

$$\frac{10}{15} = \frac{2 \times \cancel{5}}{3 \times \cancel{5}}$$

(If you don't like the idea that something is being cancelled, there is another way to write out this fraction. $2 \times 5 \div 3 \times 5$ is the same as $\frac{2}{3} \times \frac{5}{5}$ which is $\frac{2}{3} \times 1$. Done this way, you can see that nothing is magically 'cancelled', but the calculation is made much simpler).

Game: Complicated fractions

Before actually working on simplifying fractions with your child, it might help to start by complicating them! Play around with increasingly complicated ways of slicing up the pizza: each $\frac{1}{3}$ could be $\frac{2}{6}$ or $\frac{3}{9}$ or $\frac{100}{300}$ or even $\frac{1000}{3000}$. How thin would those slices be? The point of this silliness is to get your child comfortable with the idea that different fraction names can represent the same quantity. This is a difficult idea. For example, our place value system means that the digits in 36 and 12 mean very different quantities, but the same digits used in fractions — $\frac{3}{6}$ and $\frac{1}{2}$ — mean precisely the same amount. So getting children comfortable with the idea that fractions can be made to look more complicated, but actually stand for the same amount, sets the scene for thinking about whether fractions that look complicated can be expressed more simply.

Test yourself

ii) *A huge fraction to simplify*

Simplify this fraction by cancelling out numbers on the top and bottom lines:

$$\frac{49 \times 48 \times 47 \times 46 \times 45 \times 44}{6 \times 5 \times 4 \times 3 \times 2 \times 1}$$

(For a bonus, do you recognise what this fraction represents?)

Comparing *difficult* fractions

Some fractions are hard to compare. Using the sausage sharing method it wasn't possible to decide which is bigger, $\frac{3}{5}$ or $\frac{4}{7}$. In this case, one way to understand how to compare the two amounts is to use the chocolate bar method. Imagine having a chocolate bar that you can divide into fifths and into sevenths. This means that the number of chunks in the chocolate bar must be divisible by 5 and by 7. So the chocolate bar should have 5 rows and 7 columns, like this:

Each row is one fifth

$$
\begin{array}{ccccccc}
\bigcirc & \bigcirc & \bigcirc & \bigcirc & \bigcirc & \bigcirc & \bigcirc \\
\bigcirc & \bigcirc & \bigcirc & \bigcirc & \bigcirc & \bigcirc & \bigcirc \\
\bigcirc & \bigcirc & \bigcirc & \bigcirc & \bigcirc & \bigcirc & \bigcirc \\
\bigcirc & \bigcirc & \bigcirc & \bigcirc & \bigcirc & \bigcirc & \bigcirc \\
\bigcirc & \bigcirc & \bigcirc & \bigcirc & \bigcirc & \bigcirc & \bigcirc
\end{array}
$$

Each column is one seventh

There are 5×7, or 35, chunks in the bar. Now it's easy to work out how many chunks there are in $\frac{3}{5}$. This is three fifths, or three rows, which is 21 chunks. $\frac{4}{7}$ is four sevenths, four columns, which is 20 chunks. So $\frac{3}{5}$ is bigger (just!) than $\frac{4}{7}$ because $\frac{21}{35}$ is bigger than $\frac{20}{35}$.

The chocolate bar method allows you create fractions that have the same denominator (called a 'common denominator'), which makes it easy not only to compare fractions, but also to add and subtract them.

Adding fractions

You can use the chocolate bar method to add fractions together. For example, to add $\frac{3}{4} + \frac{4}{5}$, find the common denominator, which in this case is $4 \times 5 = 20$.

- $\frac{3}{4} = \dfrac{3 \times 5}{4 \times 5} = \dfrac{15}{20}$

- $\frac{4}{5} = \dfrac{4 \times 4}{5 \times 4} = \dfrac{16}{20}$

- Add these together to make

$$\frac{15}{20} + \frac{16}{20} = \frac{31}{20}$$ (You can't simplify $\frac{31}{20}$ but could write it as $1\frac{11}{20}$)

Test yourself

iii) *Chocolate bar fractions*

a. Which is larger, $\frac{2}{3}$ or $\frac{7}{11}$?
b. Add together $\frac{2}{3}$ and $\frac{7}{11}$

Where fractions crop up

Apart from the language, another reason why fractions can be confusing is that they crop up in a wide range of problems that a child might encounter. A fraction isn't just a portion of a single whole thing, such as three-quarters of a pizza.

The following situations can all sensibly have $\frac{3}{4}$ as their answer:

- Four hungry children share out three pizzas equally, how much pizza do they each eat?

- What fraction of these dots is white?

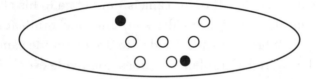

- What is the ratio of black dots to white dots?

 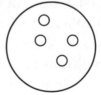

- It is 3 miles to my gran's house and 4 miles to uncle's house. What fraction of the way to my uncle's house is my gran's house?

- A baby dolphin eats 3 sprats for every 4 sprats that the mother dolphin eats. What fraction of sprats does the baby eat compared to her mother?

- Salim flips two ten-pence pieces. What is the chance that he doesn't end up with two heads showing?

- What value is the arrow pointing to on this number line?

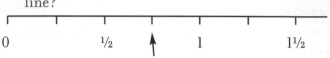

0   ½   ⬆   1   1½

Test yourself

iv) *The wise man and the camels*

An old man left his seventeen camels to his three sons, but he decided not to divide them equally. In his will, he decreed that one half of his camels were to go to his eldest son, one third to his middle son, and one ninth to his youngest. When the sons decided to divide up the camels, they discovered a problem – 17 does not divide evenly into halves, thirds or ninths. To obey their father's wishes, they were going to have to chop up some of the camels, something that they really didn't want to do. (The camels weren't too keen either.) A wise man heard of their plight. 'Don't worry,' he said, 'I have a camel that I can lend to you. You will then have 18 camels to share.' The sons were delighted, for now they could divide the camels without harming them. The eldest took his half of the camels (9), the second took his third (6) and the youngest took his ninth (2). The sons counted up: 9 + 6 + 2 = 17. There was one camel left over. 'Now you have divided your camels according to your father's instructions, I will take my own camel back,' said the wise man, leaving the three sons to scratch their heads as to how the wise man had done it. Can you figure it out?

Multiplying fractions

The idea of multiplying fractions together is well beyond the primary curriculum, but it does no harm to be familiar with what is involved. To take a cooking example, you might want half of three eighths of an ounce. The word **of** is the clue that multiplication is needed here. Whenever you hear 'one third of' or even '20 per cent of', *of* means that you will be multiplying.

How do you work out 'one third of four sevenths'? The chocolate bar enables you to see what is involved. The shaded circles represent $\frac{4}{7}$. Of these, the black circles represent $\frac{1}{3}$ of the $\frac{4}{7}$.

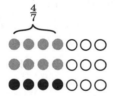

So $\frac{1}{3}$ of $\frac{4}{7}$ is $\frac{4}{21}$.

The technique for doing this for any fractions, such as $\frac{1}{3}$ x $\frac{4}{7}$ is:

- Multiply together the numerators (in this case $1 \times 4 = 4$)
- Multiply together the denominators (in this case $3 \times 7 = 21$)

Write this out as:

$$\frac{1}{3} \times \frac{4}{7} = \frac{1 \times 4}{3 \times 7} = \frac{4}{21}$$

Dividing by a fraction

What is the answer to $\frac{1}{2} \div \frac{1}{3}$?

Does this ring any vague (or loud) bells about 'turning upside down' or something? An American mum told us she was taught to remember the rule as:

'Ours is not to question why,
just invert and multiply.'

- So, $\frac{1}{2} \div \frac{1}{3} = \frac{1}{2} \times 3$ (trust us for the moment, it is).
- And $\frac{1}{2} \times 3 = \frac{3}{2} = 1\frac{1}{2}$
- So the answer to $\frac{1}{2} \div \frac{1}{3}$ is $1\frac{1}{2}$.

Yes, if you believe the rule is correct and has to be followed. But think about this for a moment. We started with $\frac{1}{2}$ and ended up with $1\frac{1}{2}$. Most people's reaction to this is something along the lines of, 'See, that's when maths starts to become meaningless'.

Let's look at this calculation in a real situation. Suppose a recipe for making pancake batter uses $\frac{1}{3}$ of a pint of milk to make one 'batch' of batter (don't worry about the eggs or flour needed). How many 'batches' of batter can you make with a pint of milk? Three.

How do you know? Because $\frac{1}{3}$ of a pint goes into one pint three times. In the same way as '2 goes into 8 four times' is a division sum ($\frac{8}{2} = 4$), $\frac{1}{3}$ goes into 1 three times is also a division: $1 \div \frac{1}{3} = 3$.

Now the answer to $\frac{1}{2} \div \frac{1}{3}$ becomes clearer:

- $1 \div \frac{1}{3}$ is 3 **thirds**.
- $\frac{1}{2} \div \frac{1}{3}$ is half of 3 thirds, in other words $1\frac{1}{2}$ **thirds**.

Decimals and Percentages

Decimals and percentages are often thought of as being somehow different from fractions like $\frac{1}{2}$ or $\frac{3}{4}$. That's partly to do with the fact that they are written differently: 0.5 or 50% look very different from $\frac{1}{2}$, but the underlying ideas are exactly the same. So why confuse matters by having different representations? If these all simply mean $\frac{1}{2}$ then why not always use that? The reason is that decimals make it much easier to compare and calculate with fractions. Remember how it was hard to compare $\frac{4}{7}$ and $\frac{2}{3}$? Using decimals and percentages it becomes much easier to compare these two fractions.

Inside kids' heads

Put these numbers in order, largest first: 0.8   0.65   0.6

Child's answer: 0.65   0.8   0.6

The child has read these numbers as being like 65, 8 and 6, and so has put 65 first. One way to get around this when talking about decimals is to put the digits into columns, and to put zeroes in the empty columns:

	Tenths	Hundredths	Thousandths
0.8	8	0	0
0.65	6	5	0
0.6	6	0	0

This links naturally with the idea of place value for tens, hundreds and thousands, and the comparison of decimals becomes much easier.

Percentage problems

Percentages are meant to make decimals even easier, particularly when comparing. They do this by making fractions into familiar, comfortable numbers between 0 and 100. Percentages have become the public face of fractions. They are used everywhere, to describe interest rates, inflation, unemployment and just about every other statistic that you can think of. Yet percentages can cause parents and children considerable headaches. Why is this?

The main problem is that percentages are introduced as 'things' in their own right. Find 20% of 160. What is 25% of 80? Calculations such as these beg the question 'why?'. Percentages are much more meaningful if they are used for the reason they were invented – to *compare* things.

Suppose Jenny gets $\frac{21}{25}$ on her French test and $\frac{16}{20}$ on her Italian. Did she do better at French or Italian? One error that children will make is to argue that she is the same on each – after all she lost 4 marks on each test. Dramatically changing the number of test questions can reveal the fallacy of this argument. Is getting $\frac{1}{10}$ right as good as getting $\frac{91}{100}$ right? Comparing the two test scores means putting them both on a common scale. Converting each to a mark out of 100 (per cent) is the commonly agreed scale to use. The fraction $\frac{21}{25}$ converts to $\frac{84}{100}$ and $\frac{16}{20}$ to $\frac{80}{100}$. Now it's clear that Jenny did better on her French test.

Inside kids' (and adults') heads

There are three things that particularly confuse children about percentages. In truth, many parents struggle with these too:

1. Percentages aren't only used as simple fractions to be added and subtracted, they are also used to describe how much things have increased or decreased. So, for example, a factory might report that prices have increased by 5%. To increase a number such as 48 by 5%, multiply 48 by $\frac{5}{100}$ (answer 2.4) and add that to the original number $(48 + 2.4 = 50.4)$.

2. '100%' is generally understood to mean 'everything'. So what does it mean to hear that something has increased by 200%, or in the case of inflation in Zimbabwe, several million per cent? (Footballers don't help when they claim to be '110% committed to the team'). In fact, percentages can be any value, once it's accepted that all per cent means is *divided by 100*, though it's common to make the slip that if something increases by 200% (or 2) that it is doubled. In fact, increasing £100 by 200% means increasing it by £200, which means multiplying £100 by *3*.

3. The biggest trap of all is the per cent button on a calculator. You'll find this discussed in the Calculator Maths chapter, on page 278.

QUICK TIP

You'll often be confronted with percentage questions in everyday life such as: 'What is a 30% discount on this price?' To make these calculations simpler, you may find it helpful to always start by working out 10%, so for example 10% of 120 is 12. Now it is easy to work out larger or smaller percentages by scaling the 10% up or down. So 5% of 120 is just half of 10%, i.e. 6, while 30% of 120 is three times 10%, i.e. 36.

Test yourself

v) *Percentages*

a. In a recent survey of 220 parents, it was found that 33 disagreed with the school's uniform policy. What percentage of parents does this represent?

b. You recently bought a toaster at full price costing £45. In the summer sales the same toaster was on offer with a 40% discount. What was the sale price?

c. A shop on your local high street is offering customers a choice of special offer. Either they will knock 10% off the basic price and then add on VAT, or they will first add the VAT and then knock 10% off the whole price. Which deal should you go for? (Call VAT 20% to make the calculation easy).

SHAPES, SYMMETRY AND ANGLES

Q. Circle the right angle on this irregular pentagon

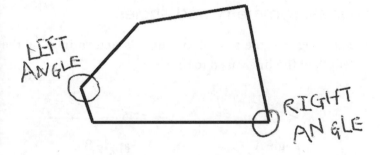

LEFT ANGLE

RIGHT ANGLE

Shapes and angles – geometry in other words – is where classical maths began, with the ancient Greeks. Shapes like triangles and pentagons can be the source of beautiful patterns, and are the basis of the visual and artistic side of mathematics. Shapes can also involve a considerable amount of reasoning and visualising, and can therefore present some real challenges to children. Mathematicians are interested not only in the shapes themselves, but also in how to describe the position of shapes in spaces. This was a critical part of the maths that enabled man to go to the moon. Your child will learn about the coordinate system at the heart of maps and graphs .

Common problems children have with shapes, symmetry and angles

1. Thinking that the size of an angle is determined by the length of the lines used to draw it:

angle A   angle B

(Here the child thinks angle B is bigger than angle A.).

2. Not realising that a square is always a rectangle (but not all rectangles are squares).

3. Thinking that hexagons always look like this:

And that this cannot be a hexagon:

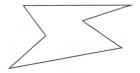

The names of the shapes

The regular shapes – triangles, squares, pentagons, and so on – have featured in mathematics since the ancient Greeks. That explains their names, with seven of the shape names coming directly from the Greek numbers:

Number of sides	Greek number	Shape name
3	Tria	Triangle
5	Pente	Pentagon
6	Hex	Hexagon
7	Hepta	Heptagon
8	Okto	Octagon
10	Deka	Decagon
12	Duodeka	Dodecagon

Confusingly, the name of the general four-sided shape, quadrilateral, came from Latin, as did the name 'square', (it's linked to the word 'quad'), while the nine and eleven-sided shapes are so rare, you'd struggle to find anybody who can tell you their names (for the record, it's either the enneagon and hendecagon or the nonagon and undecagon, depending on whether you opt for the Greek or Latin names).

Although the numbers in the names do describe the number of sides that the shape has, that isn't what 'gon' means. The 'gon' part of the name comes originally from the Greek word gonu, meaning knee, which came to be used to describe an angle because your knee makes an angle. So a hexagon is really a shape with six knees. A shape is described as 'regular' if all its sides are the same length and all the angles are the same, so for example this is a regular hexagon:

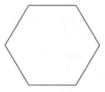

On the other hand, these two hexagons are not regular (at least in the strict mathematical sense of the word):

Shapes with more sides tend just to be referred to as polygons, which means many-sided, though you can attempt to amuse your children by saying that a polygon is also a parrot that has disappeared.

Game: i-Spy . . . a hexagon

Interesting shapes are in the house, in the streets and out on your travels. Some shapes can be found everywhere. If you look around the room you are in, you'll probably have no difficulty spotting several rectangles and a couple of circles. But other shapes are tougher. You can turn it into an I SPY game, allocating different points for different shapes. On a car journey, get your children to spot:

Road signs and roofs are the most common places to spot triangles. (Triangles are surprisingly hard to find indoors, however.) Corner stairs and the side of a regular on/off switch in a socket are two examples.) **Score: 1 point.**

Not many objects or buildings feature pentagons (the famous Pentagon in Washington DC is a rare example). You can however find them in other common objects if you know where to look. Most footballs have pentagons on them (see page 211). Cut an apple across its 'equator' – the five pips form a regular pentagon. Cut an unpeeled banana in half and you'll find the cross-section has five sides that form an irregular, and slightly curved, pentagon. Take a long thin strip of paper, tie a simple knot, and flatten the paper gently. The knot forms a regular pentagon (hold it up to the light and you might see this more clearly). **Score: 5 points.**

A beehive is made up of regular hexagons, but you don't see the inside of a beehive every day. If you look at a dice, or any other cube, tilted so that the corner is pointing towards you, the outline of the cube is a hexagon. Hexagons feature on most footballs, and many drinking glasses found in kitchens and restaurants have a hexagonal base. Most pencils are hexagonal prisms – as are modern packs of Smarties. **Score: 4 points.**

The only heptagons you are likely to see when out and about are the 20 and 50 pence coins, which are both rounded heptagons. (Coins with an odd number of sides have a constant diameter, and therefore work in slot machines because the machine can detect the edges which ever way you insert them.) **Score: 10 points.**

The width of a 50p piece remains the same whichever way you turn it

A regular STOP road sign is an octagon. Octagons combined with squares were also popular in Victorian fireplaces and tiled paths, so you might find some on the floor of a public building (or even your home if it's old). Your local bandstand is probably an octagon (it might be a hexagon though that's rare, and other bandstand shapes are extremely rare), and churches and other grand buildings often have octagonal spaces – easy to build because they are simply a square with the corners removed. Indoors, take a look at the top of screw-top cosmetic or correction-fluid bottles and you'll probably find an octagon there somewhere. **Score: 5 points.**

Shapes with more than eight sides are extremely rare. You might find them in drinking glasses or some buildings, and the occasional foreign coin: the Canadian dollar coin, known as the loonie, is a rare example of a hendecagon (that's the eleven-sided shape), while the Australian 50 cent coin, and the old British pre-decimal threepenny bit are both dodecagons (twelve sides). And at a fairground we found a hexadecagon (16 sides), the base of a carousel. One reason for the rarity of these shapes is that they become so close to circular that it's easier to make a circle than all those fiddly straight sides. **Score: 20 points.**

Tessellation – fitting shapes together

Many regular shapes fit together, to make tiled floors, mosaics, quilts and other decorative objects. This fitting together, formally known as tessellation, is not only the starting point for much geometry, but it's the area of maths that's most appealing to children with a taste for arts and crafts. You can get them doing mathematical tasks without them realising it's maths.

The most obvious tessellation is of squares and rect-angles – look at most kitchen floors, walls and pavements for examples – but things get more interesting with other shapes.

If you take *any* triangle, you can always tile a floor with identical copies of it, by (for example) joining the longest sides of the two triangles together:

The same is true of any quadrilateral, from squares . . .

. . . to trapeziums:

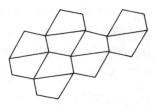

. . . to *Star Trek* badges (as long as they have straight edges):

Regular hexagons also tessellate, as any bee can tell you:

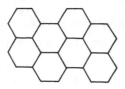

Regular pentagons don't tessellate – they leave an awkward gap:

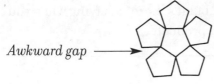

Awkward gap —————→

. . . but some *irregular* pentagons *do* tessellate. In fact, there's a simple way of finding a pentagon that will tessellate: two of the sides have to be parallel to each other. For example:

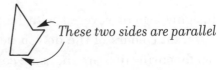

These two sides are parallel

When tessellated, this forms an interesting almost 3-D effect in a floor tiling:

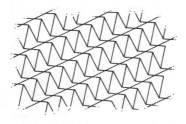

QUICK TIP

Pastry cutters are usually round, which leaves lots of waste dough or pastry that has to be re-rolled. Why not make tessellating biscuits instead? These days you can find pastry cutters that are triangular, diamond-shaped, even hexagonal. Make some hexagonal biscuits without having to waste any of the dough (except around the perimeter). You can do the same thing with plasticine, but it's more satisfying when you get to eat the finished product.

Test yourself

i) *The tiled floor*

Imagine a floor covered with hexagonal tiles. How many different colours of tile do you need to be certain that no neighbouring tiles are the same colour?

Platonic Solids

Equilateral triangles, squares and pentagons can all be joined together to make solid shapes known as the platonic solids. There are five platonic solids altogether. Three are made up from triangles:

Tetrahedron   Four triangles forming
              a pyramid

Octahedron   Eight triangles

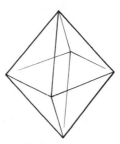

Icosahedron   Twenty triangles

('Hedron' literally means chair, but came to mean the flat face of a solid.)

The two other Platonic solids are:

Cube   Six squares

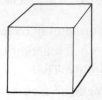

Dodecadhedron   Twelve pentagons

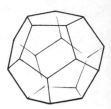

Nets

Folding flat shapes into three-dimensional objects has become a feature of primary school maths. By the age of seven your child will probably have encountered 'nets', as they are called, by opening up and flattening out cardboard boxes or by building 3D shapes from interlocking polygons. By the age of eleven they will be expected to be creating nets not only for cubes but also for prisms (such as a Toblerone® box) and other regular objects.

Here are two nets that can be folded up to make a cube:

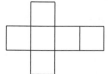   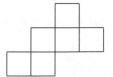

But making a net for a cube isn't just a question of sticking six squares together in any pattern. For example, this net does not make a complete cube:

Whichever way you try to fold it, there will be two over-lapping panels.

Do you find this easy to visualise? If so, you are lucky. Most children (and many parents) can only work out if a net will fold up to make a solid object by physically folding the pieces, at which point the penny usually drops. In the class-room they will typically be given the opportunity – and the time – to fold up their own nets. In school tests, however, they will be expected to do this in their heads. The only way to learn it is by practice.

QUICK TIP

You can begin to explore nets at home by cutting a cereal box into the six rectangular faces that make it up and exploring different ways of taping the pieces back together to make a net that folds up to recreate the box. There are a surprising number of ways of doing this.

There are people who design nets for a living, for example the designers of the 3-D objects printed on the back of cereal packets (the type where you have to tuck tab A into slit B). You can follow their example, and make nets into a fun, creative activity for the home. You can create your own dice out of sheets of cardboard, or make a fold-up treasure chest (complete with its curved top). But why not push the boat out and make something spectacular with your child? Would you believe that the following net made up of tri-angles folds up to make an (icosahedral) model of Planet Earth? Every home should have its own globe, but how much cooler if your globe is folded up from a net? (Nets like this are easily found on the Web.)

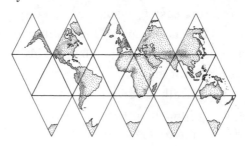

Test yourself

ii) *A strange net*

What will this look like when it is folded into a solid object?

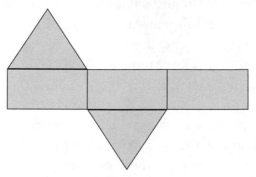

Footballs and hexagons

Everyone knows what a football is like — you know the one, with black and white shapes all over it, that you used to be able to buy for a pound at Woolworths. But can you draw one without looking?

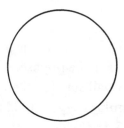

Most parents who make an attempt end up with something like this:

Which looks . . . wrong. The reason is that most people imagine that a football is composed entirely of hexagons. Illustrators, cartoonists, even the civil servant who designed the standard football stadium road sign to be seen across the UK, make the same mistake. (Take a look at any road sign for a football club, and you'll see what we mean!) But regular hexagons only fit together when they are flat, and if you attempt to make a football out of hexagons, you will end up with lumps and creases.

Now take a look at a real football:

It is, in fact, made up of hexagons and pentagons – 20 hexagons and 12 pentagons, to be precise.

You can make a football out of an icosahedron. The icosahedron has 20 triangular faces, with 12 corners (or *vertices*) each of which has 5 triangles meeting at it. If you snip off a corner you form a pentagon.

Snipping off all 12 corners leaves you with a familiar-looking shape.

Because of this snipping off, the resulting solid is known as a *truncated icosahedron* – but most people know it simply as a football.

Game: i-Spy 3-dimensional shapes

You can extend the earlier I-Spy game to looking out for interesting regular three-dimensional shapes. Shapes that involve circles are the most common: spheres and cylinders are everywhere (in the form of balls, pipes, food packaging, broom handles and so on). Cubes are commonplace (particularly small ones, in the form of sugar lumps and dice) and 'triangular prisms' form most roof spaces, as well as Toblerone bars. But pyramids are quite hard to find (PG Tips teabags aside), and more exotic shapes such as dodecahedrons are very rare.

Some other common 3-D shapes

Sphere

Cylinder

Cuboid

Square-based pyramid

Cone

ANGLES

One of the earliest reasons why mankind wanted to measure angles was in the study of the stars. It was the Greeks who decided that the way to mark out a circle was to divide it into 360 'degrees'. Why 360 and not 100? Nobody is certain, but there is one very plausible explanation: of the numbers from two to twelve, 360 divides exactly by 2, 3, 4, 5, 6, 8, 9, 10 and 12, without leaving any fiddly remainders or fractions. In comparison, 100 divides only by 2, 4, 5 and 10. Since fractions were not popular in Greek times it helped to have a number that could be divided up so easily.

360 is also very nearly the number of days in the year, so in the cycle of life it would make sense to mark out the cycle with a nice round number to represent a calendar.

If 360 degrees represents a full circle, then turning through a semi (half) circle must be 180 degrees:

Right, acute and obtuse angles

A a quarter circle is 90 degrees, generally known as a right angle, and indicated with straight lines forming a square in the corner:

(The word 'right' here is linked to the word 'upright' and is nothing to do with being either 'correct' or 'on the right-hand side' – children often assume there's also something called a 'wrong' angle or a 'left angle'.)

Angles less than 90 degrees are called 'acute' and those greater than 90 and less than 180 degrees are called obtuse. (For completeness, an angle greater than 180 degrees is called a reflex angle – though you'd struggle to find many people who know this.)

Game: Right-angle treasure hunt

You can organise a right-angle treasure hunt at home by making a sturdy right-angle 'tester' out of a sheet of scrap paper. Simply take a piece of paper and fold it:

Now make a second fold that makes the first fold exactly meet itself.

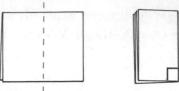

The corner created where the fold meets itself is a pretty accurate right angle. Use this to test whether objects that look like they have right angled corners really do.

Triangles

Triangles come in three types, and your child will be expected to become familiar with all of them:

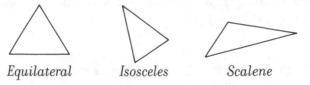

Equilateral   *Isosceles*   *Scalene*

- Equilateral triangles have three sides (and angles) that are the same.
- Isosceles triangles have two sides (and angles) that are the same.
- Scalene triangles have three different sides (and three different angles).

One curiosity of triangles is that if you take any triangle and measure its three internal angles using a protractor, they will always add to 180 degrees. An easy way to demonstrate this to yourself is to draw a triangle on a piece of paper, tear off the three corners, and put them together.

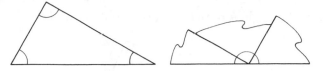

The three angles make a straight line which is 180 degrees.

This means that if you know any two of the angles in a triangle, you can work out what the third is. For example, if two of the angles in a triangle are 30 degrees and 80 degrees, the third must be 180 − (30 + 80), which is 70. And this little building block of knowledge turns out to be fundamental to vast amounts of more sophisticated geometry in higher maths − which is one reason why your children learn it.

Test yourself

iii) *The right-angled triangle*

Sarah has drawn a triangle, and one of its angles is a right angle. Which of the following is a possible shape of the triangle?

a. Equilateral
b. Isosceles
c. Scalene

Test yourself

iv) *Parked car*

What is the angle A between the car and the wall?

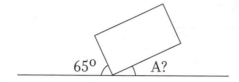

SYMMETRY

The natural world is full of symmetry, and mathematics is too, and since the basic ideas are easy to grasp it's no surprise that symmetry features prominently in primary school maths. The two main ideas of symmetry that your child will encounter are:

- Reflective symmetry, in which half of a shape is the mirror image of the other half.
- Rotational symmetry, in which a shape is identical after a partial turn.

The regular shapes described earlier in the chapter all have as many symmetries as they have sides, so for example a square has four lines of (reflective) symmetry, and four rotational symmetries:

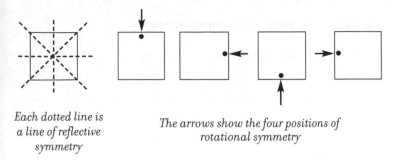

*Each dotted line is
a line of reflective
symmetry*

*The arrows show the four positions of
rotational symmetry*

Meanwhile, a regular pentagon has five symmetries, a hexagon has six, and so on.

Test yourself

v) *Cocktail sticks*

Five cocktail sticks have been placed on a table to make an 'antelope'. *Remove* one cocktail stick to leave a shape that has a line of symmetry.

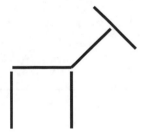

Reflecting and rotating shapes

Children are expected to rotate and reflect shapes too – not just by sketching, but by accurately plotting the position of a shape when it has been reflected. This is done using 'cartesian co-ordinates', the system invented by Descartes (the same man who famously declared 'I think, therefore I am').

Co-ordinates are very simple: the position of a point on the grid is indicated by saying how far across and up it is, using the scales on the side and the bottom.

The only confusion comes in remembering which number comes first, up or across? In fact, the convention is to state *across* first and then *up*. To remember this, think of going into a house: first you go along the hallway, then up the stairs. So the position of point X, below, is given as 5 along and 3 up, which is written as (5,3).

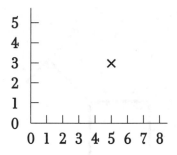

Game: Battleships

One of the best ways of reinforcing the idea of co-ordinates is to play the game Battleships. This game can be played on squared paper, using, say, a 10-by-10 grid. Each player shades in a number of different shapes at secret locations around the grid (the shapes represent, for example, an aircraft carrier, which is four squares in a line, and a battleship which is three in a line) and the players then take it in turn to call out co-ordinates on the other player's grid. If the co-ordinate is on one of the opponent's ships, it counts as a hit, and a player wins when all parts of every one of the opponent's ships have been hit.

Test yourself

vi) *Where is the square?*

Mark and write down the co-ordinates of two more points that joined to the co-ordinates marked here would make a square. Can you find three ways of doing this?

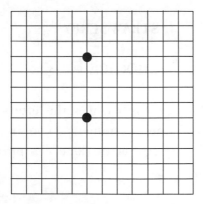

Mirror tricks to try at home

(1) The ice cream mirror trick

Ask your child to name their favourite three ice creams (vanilla, raspberry and strawberry, say). Write them down in capital letters on a sheet of paper, and draw pictures of each one in a cone. Then tell her that your favourite is CHOC ICE, and write that down with a picture. Now ask: 'What is so special about a CHOC ICE?' There will be lots of suggestions (for example 'it has chocolate in it' and 'it has a different shape'). Now announce that what's really special about Choc Ice is that it doesn't get ruined when you turn it upside down. Turn the page below upside down, and look at it in a mirror. The upside down cones will be ruined of course, but now look at the words. All of the words are 'ruined' too, except for CHOC ICE, which miraculously still appears as CHOC ICE.

VANILLA

RASPBERRY

STRAWBERRY

CHOC ICE

(The reason this works is that all of the letters in CHOC ICE have a horizontal line of symmetry. Looking at the words upside down in a mirror is the same as flipping each letter top to bottom.)

(2) The decoded riddle

A similar idea to the Choc Ice, but this time presented as a riddle. Write out the following paragraph on a sheet of paper (in block capitals).

> THIS IS A MYSTERIOUS PARAGRAPH. IT CONTAINS A SPECIAL WORD THAT IS GOING TO BECOME DECODED WHEN YOU TURN THE PARAGRAPH UPSIDE DOWN AND LOOK AT IT IN THE MIRROR. ALL OF THE OTHER WORDS WILL BE SPOILED. CAN YOU SPOT THE MAGIC WORD?

(3) The way out sign

On a clear sheet of plastic, write the words:

W
A
Y

O
U
T

Say: 'This is the door of the room where you never escape. What is on the other side of the door?' When you turn the door over to look at the other side it says . . . WAY OUT.

Palindromes and number symmetry

Another form of symmetry is the 'palindrome', something that reads the same forwards and backwards. Girls called HANNAH and ANNA are always pleased to discover that their names are palindromes, as are boys called BOB and OTTO. There's a lot of fun to be had with finding other palindromic words and sentences, including exotic ones such as, 'Ten alps bordered Rob's planet'.

Palindromes apply to numbers too. There have been two palindromic years in the relatively recent past (1991 and 2002), though it's unlikely your children will encounter the next one in 2112. Still, plenty of other palindromic patterns can be found in numbers, including dates (such as the twenty-first of the first nine months in 2012, for example 21 March, written 21.3.12.)

If you multiply numbers made up entirely of 1s by themselves, the result is a palindrome:

$$11 \times 11 = 121$$
$$111 \times 111 = 12321$$
$$1111 \times 1111 = 1234321$$

If you have the patience to do long multiplication, or your calculator can take it, you can check this works all the way up to 111,111,111 × 111,111,111 = 12345768987654321.

MEASURING

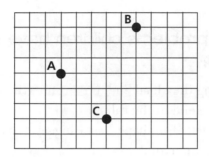

Q. A boat went from A to B.

To do this it was told: 'Go 5km east, then 3km north.'

The boat now has to go from B to C.

Write down what to tell the boat.

Go _Ready steady_ then _go_

Time, distance, area, volume, speed, weight – all these things need to be measured, and children spend plenty of school time learning about clocks, rulers, scales, and the other devices for putting numbers to measurements. Children are also, however, expected to learn to measure things using their imagination – by estimating.

Common problems children have with measuring

1. Not noticing the 'dead end' on a ruler (the part between zero and the end of the ruler).
2. Believing that if a shape has a bigger area that it must have a bigger perimeter.
3. Interpreting scales when there isn't a mark for each number.
4. Using an inappropriate scale for measuring a distance (for examples kilograms to weigh a feather).

Imperial and metric

We might live in a metric age, but in the UK we still hold on to a few treasured imperial measurements. Milk and beer are still served in pints, journeys are measured in miles, cricket pitches are still measured in yards, race courses in furlongs. Opinion is divided on whether it's time to drop these last bastions and turn entirely to the simpler metric system, though one thing to bear in mind is that the USA, perhaps the most influential culture on earth, remains firmly rooted in imperial. Engineers in the USA still work in miles, feet, pounds and Fahrenheit. So perhaps it's no bad thing that the UK remains at least partly 'bilingual' in its measurement – it is, after all, one way in which the UK and USA do still retain a special relationship! One consequence of having to know both types of measurement is that it provides the maths teacher with ready, everyday examples for working out ratios, by converting pounds to kilograms and pints to litres.

Your child will be taught almost exclusively in metric, which has the advantage that everything works in base ten, but does mean that you might inadvertently have a communication problem. If you say something is twenty yards away, your child might not have a clue what you are talking about. So at home, if you're comfortable with feet and inches, by all means use them as your first language, but try to give their metric equivalents too.

The mums and dads metrication test

Depending on your age and upbringing, you will be somewhere in the range of being mainly imperial to mainly metric. A test of how imperial or metric you are is to think about what units you would use to describe the following (do them quickly, to test your instinct!):

- the height of a mountain
- the weight of a dozen apples
- the amount of water that a bucket will hold
- the speed of a train

Score 1 point each time a metric measurement came up first. If you scored 4, you are fully metricated, and will probably be able converse fluently with your child. If you scored zero, there may be a language barrier.

Conversion table

Many of the conversions from imperial to metric can be done (as a very rough guide) just by doubling or halving. Here's a table showing the most common conversions you'll need, with the quick and dirty rough method, followed by a more accurate method:

Conversion	Very roughly	More accurate
Inches to centimetres	Double	Multiply by 2.5
Yards to metres	The same!	*'A metre measures three foot three, it's longer than a yard, you see.'* Taking off 10% to turn yards to metres is close enough for most everyday situations
Miles to kilometres	Double	Multiply by 8 and divide by 5
Fahrenheit to Centigrade	Take away 30, then divide by 2	Take away 32, then multiply by 5 and divide by 9
Pints to litres	Halve	*'A litre of water's A pint and three-quarters'*
Gallons to litres	Double twice	Multiply by 4.5
Pounds to kilograms	Halve	*'Two and a quarter pounds of jam weigh about a kilogram'*

Using the right scale

You wouldn't use millimetres to measure the distance from London to Glasgow; nor would you use kilograms to measure the weight of a feather. Why not? Because our senses are better at relating to whole numbers between one and a thousand, rather than numbers that are tiny or huge. The idea of small measures for small things and large measures for large things will make sense to your child, and you should try to talk about these measurements as much as possible.

These days, only two prefixes are commonly used for metres, grams and litres to turn them into appropriate measures:

Kilo	1,000
Milli	1/1000th

There are prefixes for the other decimals between these numbers, but (with the exception of centimetres) these are hardly used at all:

Hecto	100
Deca	10
Deci	1/10
Centi	1/100

TIME

Your child will become familiar with the idea of time (now, later, tomorrow, yesterday) as early as three, and will understand that clocks help to tell the time at a similar age. But they may not become confident with telling the time until much later – many still struggle with anything other than the 'o'clocks' at the age of eight.

Clocks are an interesting mathematical topic, since they connect many parts of maths. Although we're including them under measuring, discussion of clocks could appear under counting, addition and subtraction, fractions (quarters and halves) and the 5 times table (minutes past the hour can be found by multiplying the clockface number by 5).

Clocks are also connected to angles, because the minutes of the hour have their basis in the ancient system of counting in 60s, just as degrees in a circle do. The twelve hour marks are equally spaced around the perimeter of the clock, and so the angle between the two hands at one o'clock is one twelfth of a turn, 360/12, or 30 degrees. The hour hand points at 45 degrees to the vertical when it is midway between 1 and 2, in other words at half past one.

Test yourself

i) *Clock puzzle*

Between noon and midnight, how many times is the angle between the minute hand and the hour hand 90 degrees? (You can probably find two very quickly – but can you find a third . . . and are there more?)

Apart from learning to tell the time, the other thing about time that children find particularly tricky is adding time on a digital clock. For example, if a train leaves at 3:25 and a journey takes 1 hour and 40 minutes, what time does the train arrive? The time 3:25 looks very similar to a decimal number, and in decimals 3.25 plus 1.40 equals 4.65. In clock time, however, minutes work in base 60 not base 100, so when the minutes reach 60 you add one to the hours and return to zero. As adults we are used to this, so we tend to forget that it can confuse children.

Test yourself

ii) *Baking the cake*

Jamie is baking a cake. The time on the clock when Jamie puts the cake in the oven is 4.40 p.m. The cake needs to bake for 90 minutes. What time will Jamie take the cake out of the oven?

Clocks and direction

Clocks and direction fit naturally together. The bearings of north, south, east and west can be represented as noon, three, six and nine o'clock, and when turning to change direction we naturally use the expressions 'clockwise' and 'anticlockwise'. You can introduce the natural link between clocks and compass bearings when pointing out landmarks to your child. 'If you imagine the church is at 12 o'clock, then the hill I'm looking at is at 1 o'clock.'

Using your watch to take a bearing

Who needs satellite navigation when you can use your watch? Next time you are out and about with your family (whether out on a country walk or stuck in the suburbs), use your watch to find your bearings. Hold your watch flat, and point the hour hand in the direction of the sun. Now split the angle between the hour hand and 12 on your watch, and that is roughly south. (For more accurate results, your watch needs to be set on Greenwich Mean Time – and of course it does need to be sunny!)

LENGTH

Children need to learn about how to make precise measurements using a ruler, but they also need to have a broader feel for distance. As well as confusion over inches versus centimetres, there are two common slip-ups with rulers. Here's a broken ruler being used to measure the length of a line:

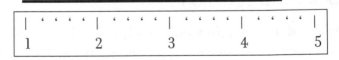

As is normally the case with rulers, there are divisions marked between the numbers that are not labelled. The user is expected to be able to interpret what these divisions represent. In this case, there are five divisions between each whole number, so each division represents 0.2, and the length of the line is 3.6 (that's 4.6 minus 1.0).

Inside kids' heads

Can you see why children measuring the line opposite would come up with these common wrong answers?

4.6

4.3

3.3

In the first case, the child was concentrating on the higher number, 4.6, and didn't think to check that the other end was at position zero. Using a ruler often requires simple subtraction like this. (You can introduce this idea by deliberately using a broken ruler that doesn't have a zero and asking, 'Oh dear, how are we going to measure the length?')

The second and third wrong answers arose because the children read the right-hand end of the line as 4.3, having counted three divisions and assumed that each division must be 0.1. You can point out why the divisions can't be 0.1 by counting along 4.1, 4.2, 4.3 . . . and realising that the fifth division – supposedly 4.5 – coincides with the whole number 5.

QUICK TIP

It's standard practice in many families to allocate some door frame or piece of wall to monitor the children's height as they grow. Children typically enjoy discovering their new height, and you can make this into a permanent feature by sticking an old tape measure to the wall so that once you have made a pencil mark next to the tape, they can easily check their heights themselves. A measure that has feet, inches and centimetres marked is great, for while children will naturally be taught to think of height in centimetres, the landmarks that we talk about for height are still four feet, five feet and — what's generally deemed to make a man 'tall' — six feet.

Other measuring scales

Measuring scales appear all over the place, on weighing scales, measuring jugs, thermometers, to name but a few. The numbering on the scale might be in regular ones, but just as often the gap between numbers will be in hundreds or small decimals. Whatever the scale, the idea behind reading it is the same — learning to interpret the divisions between the numbers, and realising that one division does not always represent one unit.

The more familiar your child becomes with reading scales around the house, the easier they will find it to interpret a new scale when it comes along.

Test yourself

iii) *Time line*

Here is part of a time line. Mark each event with an arrow to the appropriate part of the time line.

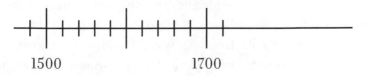

- Henry VIII becomes King   1509
- Fire of London   1666
- The English defeat the Spanish Armada   1588

PERIMETER

Perimeter comes from the Greek word 'peri' meaning 'around'. Children get confused between measuring the perimeter of something and measuring its area. You can easily get rid of the confusion by always associating perimeter with a piece of string: how long a piece of string would you need to have to go around this circle, or this swimming pool?

The perimeter of a circle has a special name, the circumference. The length of the circumference of a circle divided by the circle's diameter is always the same: a number that is a little larger than 3, very close to 3.14, and universally referred to as pi (or 'π'). Pi is not something that children become familiar with until secondary school, but you can introduce the mystery of pi by finding a range of circular items around the house – a tin of baked beans, a DVD, a pizza – and using a piece of string to verify that three lengths of the diameter is almost but not quite enough to cover the circumference of each circle.

Three diameter lengths of string nearly surrounds the circle.

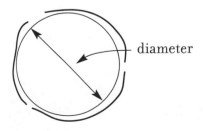

diameter

Test yourself

iv) *Perimeter*

This shape is made from six squares.

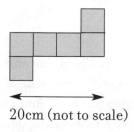

←——————→
20cm (not to scale)

Calculate the perimeter of the shape.

AREAS AND RECTANGLES

The easiest way to think about area is as a flat space. Areas can be covered by carpet, or grass, wallpaper or paint, and children can understand that areas might need to be measured in order to check that you have enough carpet, paint, etc., to cover them.

The first step children use to measure area is by counting squares. By drawing a shape on squared paper, it's easy (if a bit laborious) to count up how many squares are enclosed by the shape. The easiest area to measure is for a rectangle.

```
    1  2  3  4  5  6  7
  ┌──┬──┬──┬──┬──┬──┬──┐
1 │  │  │  │  │  │  │  │
  ├──┼──┼──┼──┼──┼──┼──┤
2 │  │  │  │  │  │  │  │
  ├──┼──┼──┼──┼──┼──┼──┤
3 │  │  │  │  │  │  │  │
  └──┴──┴──┴──┴──┴──┴──┘
```

Here, a careful count will confirm that the number of squares is 21. But the penny will quickly drop that there's a quicker way. There are 3 rows of 7 squares, or 7 columns of 3 squares, either way that's 3 × 7. In other words, there's no need to count squares at all, the area of a rectangle can be found by multiplying the lengths of the two sides.

Areas of more complicated shapes

The counting squares technique works for every other shape, too, though there won't always be convenient whole squares to count. For example:

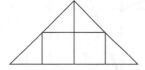

In this triangle there are two whole squares and four half squares, making four squares in total.

Primary school children are rarely required to count anything less friendly than half-squares, but the idea can be taken further. The rule for making a very good estimate is to count any part-squares that are more than half as *whole squares*, and any that are less than half as *no square*. So in the triangle below there are 6 whole squares, and 3 more-than-half-squares, giving an estimate of 9 squares.

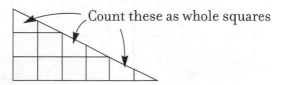

Count these as whole squares

Test yourself

v) *The mysterious appearing square*

Here is a fiendish puzzle that appears to make a square appear out of nowhere! This is an ordinary grid, with 8 squares across and 8 squares down making 64 squares in total.

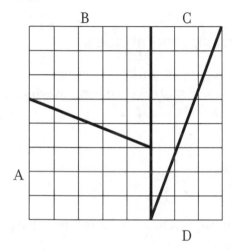

As you can see, the grid has been cut into four pieces, A, B, C and D. These exact same pieces can now be put together to make a rectangle like this:

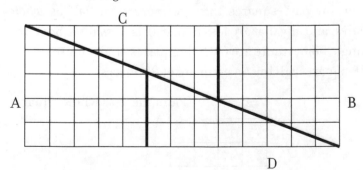

Check out that all the pieces are the same shape as in the original. Now count up the squares. The rectangle has 13 squares across and 5 down the side. 13 times 5 is 65 . . . but that is one more than in the original. Where has the new square come from?

VOLUME

Primary school children won't be expected to learn how to calculate the volume of a solid object, but they will learn about measuring the volume of liquids using litres, and they will be expected to have a grasp of when there is more or less than a litre of water in a jug.

Even though they won't be looking at volumes in any detail, it does no harm to introduce them to some of the ideas connected to it. One interesting experiment is to take two sheets of A4.

Roll one along the long edge to make a long thin tube, and roll the other along the shorter edge to make a more squat tube:

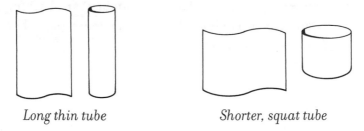

Long thin tube   *Shorter, squat tube*

These two tubes are different shapes but made out of the same piece of paper. Which one will hold the most dried pasta?

If your instinct is that they should hold the same amount, then your instinct is wrong. The short squat tube holds more – in fact about 40 per cent more.

Now cut the A4 sheet in half and make two identical smaller cylinders. Which holds more volume, the larger one or the two small copies? First instinct might be that it makes no difference, but when you make the cylinders the answer becomes apparent.

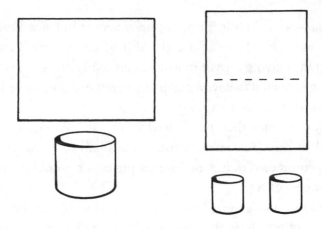

Just by looking at them you can see that the two small tubes will contain less than the large one. In fact, once again, the large tube contains about 40 per cent more than the two smaller tubes combined.

WEIGHT

Your children will be introduced to grams and kilograms at school, and will be taught about using scales to measure weight. Not every school has old-fashioned balancing scales, which is a shame since these can be great for building an understanding not only of weight, but also of fractions and even the idea of 'equations' (an equation is just like a balance, where the amount on one side is the same as the amount on the other side).

You have to hunt around for them these days, but there was a time when every kitchen would have had a set of scales in which ingredients were placed in a pan that was then balanced with weights on the other arm. Maybe Grandma still has a set in the back of the cupboard. The weights would normally be the following (in ounces):

$\frac{1}{4}$ $\frac{1}{2}$ 1   2   4   8   16 (= one pound)

Notice how as they increase, the weights double each time. What was clever about this old imperial system was that with these seven weights you could measure everything between $\frac{1}{4}$ ounce and 31 ounces down to the nearest $\frac{1}{4}$ ounce. For example, $27\frac{1}{4}$ ounces is $16 + 8 + 2 + 1 + \frac{1}{4}$. This idea of doubling each time is known as the binary system, and it happens to be the most efficient way of measuring weights using a normal set of scales. (We also met binary numbers on page 63 – there are lots of connections in mathematics that it helps to point out to your child.)

Today, the equivalent scales use metric weights. 16 ounces is not far off 500 grams, and because we like numbers in round hundreds, the weights you'll find today are usually these:

	500g
2 ×	200g
	100g
	50g
2 ×	20g
	10g
	5g

This allows you to make up every weight from 5g to 1,105g to the nearest five grams, a similar range and accuracy to the imperial weights but needing nine weights instead of the old seven.

If we adults were prepared not to have hundreds, then to cover the same range as the old imperial system, we could use eight metric weights instead of nine, for example:

$$5 \quad 10 \quad 20 \quad 40 \quad 80 \quad 160 \quad 320 \quad 640 \ldots$$

These eight weights allow you to weigh everything up to 1.275 kg to the nearest 5 grams.

Next time you're visiting the grandparents, see if they have a set of imperial scales – and get your children to help measure the ingredients for a cake using the efficiency of binary maths.

Measuring benchmarks

One helpful thing to work on with your child is developing a set of measurement 'benchmarks' – informal everyday units of measure that you can use to think about objects and to estimate measures when no formal tools are around. Here's a starting list – talk with your child about others that could extend it:

250 grams:	a packet of butter
1 kilogram:	bag of sugar
1 litre:	large carton of milk
200 millilitres:	yogurt carton
15 ml:	medicine spoon
1 metre:	dad's long stride
30cm:	dad's shoe (one foot)
25cm:	mother's shoe
2 metres:	length of bed
2 minutes:	time it takes to brush your teeth (properly!)

DATA-HANDLING
AND CHANCE

Q. Circle the group of 5.

This little girl liked cats, so she decided to adapt the question.

W hen many parents were at school, statistics and probability barely featured in school maths. These days not only have they become a significant part of the curriculum, but both topics are introduced at primary school – though they go by the names Data-handling and Chance.

Data-handling is a dry name for something that children can find very engaging: posing a question, deciding on the information that they need to collect in order to answer the question, compiling, sorting and presenting this information into charts, interpreting those charts and using that information to understand the world around them. Meanwhile any child that plays card or dice games will be familiar with the idea of chance (or probability), though at primary age children are likely to think of it rather less scientifically as 'luck'.

Common problems children have with data-handling and chance

1. Thinking that 'negative' information is less helpful than positive. For example, playing 'guess my number' – a version of 20 questions where one player has to think of a number and the other has to establish what the number is by asking questions with yes/no answers. Asking 'is the number even' and being told 'no' conveys just as much information as asking 'is the number odd' and being told 'yes' but children interpret the 'no' answer as less helpful.
2. Thinking that a graph provides a 'picture' of an event and interpreting it incorrectly as a result.

3. Not realising that pie charts look at relative proportions of a group, rather than the size of a group.

Tallying and frequencies

If you've ever used the familiar bar gate tally system – H̶H̶ – to keep track of something, say, scores in a game or people entering the school hall, then you are joining in a long tradition of such activity. Around 40,000 years of tradition! We know this because of the Lebombo bone, which is the oldest known mathematical artefact. It's a baboon bone that some caveperson made tallymarks on. It's dated to be around 37,000 years old and presumably isn't the first such bone. Of course we've no idea what was being tallied – days elapsed, animals caught, members of the tribe – who knows? But whatever it was, our species has a long history of wanting to keep track of quantities.

Early examples of tallies have no structure to them but grouping the tallies in some way is evident in bones that were found later, and grouping in fives eventually became the norm.

The tally system is the simplest way of collecting and presenting data. Keeping tallies of two things – the boys' team score versus the girls', for example – is really the same as a horizontal bar chart. In both cases, you tell instantly from looking at the tallies, the relative scores of the two things you are counting.

Sometimes tallies are used simply to record one type of thing – the number of people attending a show, the number of vehicles going into the car park. But charts can be drawn up to record the tallies under different headings; types of vehicles going past, favourite flavours of crisps, colours of Smarties in a tube. Such tally charts then give a sense of how often certain events occur: the commonest vehicle, the most popular flavour, how many Smarties are blue compared to red. These are known as frequency charts and are often then presented in graphs to make the data visual.

Comparing, sorting, organising

Comparing, sorting, and organising are at the heart of data-handling. Look at the number of ways in earlier chapters that we sorted and named numbers — whole numbers, fractions, odd/even, square, multiples, factors, and so on. Or the myriad ways of sorting and naming shapes — symmetrical, two-dimensional or three-dimensional, regular, and so on.

At the heart of data-handling is the skill of spotting what features things have in common, and what their differences are, and then sorting them into categories.

Odd one out?

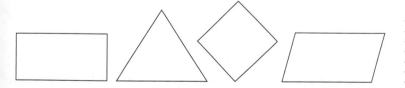

Which of these four shapes would you say is the odd one out?

Most people immediately choose the triangle, because it is the only shape with three sides. That's one possible reason for it being the odd one out. Before reading on, can you think of any other reasons why the triangle is the odd one out? Can you come up with at least five reasons?

Here are some:

- Its angles add up to 180 degrees (the others all add up to 360)
- It has three lines of symmetry
- It has the smallest area
- It has the smallest perimeter
- It has three angles
- All its angles are acute
- Six of these triangles will fit together to make a hexagon

The triangle is not the only possible odd one out. Each of the other shapes has something that makes it distinctive – all but one have equal sides, all but one have a flat base, all but one have equal angles – which demonstrates that categorising is rarely a cut-and-dried exercise. Since data-handling involves putting information into categories, the more experience children have of thinking about and creating categories the better.

Game: Odd one out

'Odd one out?' can become a creative game to play with your child at any age. With young children you can explore a collection of household objects. Collect together, say, a spoon, a cup, a bottle and a fork. Can you and your child find reasons why each of these might be the odd one out? What about a pencil, a wooden ruler, a felt tip and an eraser? Who can find the most unusual or silly reason why one thing is the odd one out? To get you going, here are a few:

- A pencil is the odd one out because it's the only one that doesn't have a bendy part
- The ruler is the only odd one out because it's the only one that makes a silly noise when you set it vibrating of the end of a table
- An eraser is the only one that has another name (a rubber)

Older children can explore sets of numbers. How many different answers are there to the question of which is the odd one out of: 20, 15, 24, 25?

A variation on 'odd one out' is 'What am I thinking of?'. Decide on a collection of things to choose from. These may be items around the home that your child can see, for example, clothes in the washing basket, or items on the table. Or mathematical things to imagine: a whole number less than 100, a 3-D shape. Can your child figure out what you are thinking of by asking questions with yes/no answers? For example, there is a cup, a glass, a sandwich and an apple on the table. A round of 'What am I thinking of' might go something like:

Can you eat it?
No.
Is it see-through?
No.
Is it the cup?
Yes.

Game: Find the number

You and your child take it in turns to think of a number between 1 and 100, and then ask questions with yes/no answers to figure out the number. How quickly can you figure it out?

> I'm thinking of a number between 1 and 100.
> Is it even?
> Yes.
> Is it greater than 50?
> No.
> Is it a multiple of 3?
> Yes.
> Is it greater than 25?
> Yes.

And so on.

The best strategy in games like this (that is, the strategy that will home in on the right answer the fastest) is to ask questions that split the remaining possible answers into two equal groups, half of them 'yes' and the other half 'no'. So asking 'is the number even?' is a great first question because half the possible numbers are even, and the other half aren't. On the other hand, asking 'does the number end in zero?' is not a good question, because 90 of the 100 numbers don't end in zero, so there's a high chance of ending up with the answer 'no'.

Sorting with Venn diagrams

At some point in their primary schooling, your child will be introduced to two ways of sorting — using a Venn diagram (which you've probably heard of) and a Carroll diagram (which you probably haven't). Although Venn and Carroll diagrams essentially do the same thing, they present data in slightly different ways and so emphasise different things.

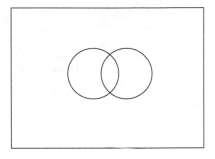

Venn diagram

You can play with Venn diagrams with your child by simply making two loops of string and sorting a collection of objects. For example, you might collect together an apple, an orange, an orange wax crayon and a pencil. Explore ways of splitting the collection in two, for example, things to eat and things to write with. With a choice between categories such as 'things to eat' and 'orange', it won't be long before your child realises that some objects need to be put into both sets at once: the orange into 'things to eat' and 'orange'. The two loops of string need to be overlapped.

It may not be immediately obvious that there is a fourth place that objects can belong – the space around the loops.

In this example, 'things you cannot eat and are not orange' will be outside the loops.

At primary school, Venn diagrams are often only presented as two intersecting loops, but in some situations one loop can be completely inside the other. For example if one loop represents 'animals that make milk' and the other loop represents 'cows' then since all cows make milk, the cow loop can go inside the milk loop, to make a Venn diagram that resembles a fried egg:

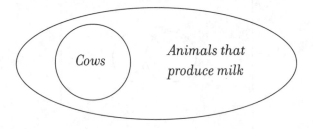

And there's no rule that says there should be only two loops. You can have as many loops as you want to represent different categories, though the intersections become complicated for more than three.

Carroll diagrams

The 'Carroll diagram' is named after Lewis Carroll, who is best known for writing *Alice in Wonderland* but was also a maths lecturer at Oxford. Lewis Carroll (real name Charles Dodgson) had a particular interest in logic, and in 1896 he wrote a book on this subject which featured what he called a 'Biliteral diagram'. It has since been renamed the Carroll diagram, and every primary school child is now taught it as a method of categorising objects.

The Carroll diagram lends itself particularly to being used to sort mathematical objects. For example, number cards can be sorted on a number of properties: odd/even, greater than/less than 20, is/is not a multiple of five . . .

	ODD	EVEN
MORE THAN 10	13, 17, 29	16, 88, 52
LESS THAN 10	1, 3, 7, 9	2, 6, 8

Test yourself

i) *Carroll diagram*

Where would the numbers in the Venn diagram be placed on the Carroll diagram?

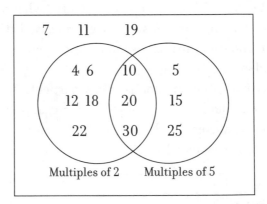

	MULTIPLE OF 2	NOT A MULTIPLE OF 2
MULTIPLE OF 5		
NOT A MULTIPLE OF 5		

Data handling projects

Not too long ago, most children's experience of data-handling at school largely consisted of being given a set of numbers and drawing bar graphs. They spent a lot of time colouring in, but they did not learn that much about data handling. Now producing the graphs can more quickly be done by technology – a spreadsheet can quickly transform data into a graph or chart – so more attention can be paid to a complete data-handling project.

This starts with choosing a question to investigate. It could be something very basic: 'Are there more yellow cars than green cars on the main road?', which is simply a matter of tallying and then presenting a comparison of the numbers. But children can take on more subtle investigations in class, for example: *Does practice improve learning your times tables?*

In class they would then discuss *how* they could explore this. For example, they might agree on a particular table that they will practise for, say, a week (for example the 3 times table). They would gather data on how the class was performing on the 3 times table at the beginning of the week, by organising a tables test for everyone. They would practise the table over several days and measure everyone's performance again at the end of the week.

Presenting results

There are many different ways to present results. These include bar charts, line graphs, pie charts, but many more subtle forms too, such as scattergraphs. One of the decisions to make in a data-handling project is which type of chart to use to display the results, as different charts allow you to draw different conclusions.

For the question 'Does practice improve learning your times tables?', younger children might decide to plot each pupil's score side by side on a block graph:

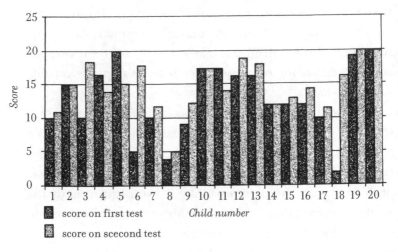

They would then interpret their graphs by asking questions such as:

- What was the greatest improvement?
- Did anyone perform worse on the second test?
- Did everyone improve? How many did improve?
- Overall, did the class improve?

Older children might choose to use a scattergraph which can be more revealing. In a scattergraph, each child's score in the first test is plotted against their score in the second test. The diagonal line on the scattergraph marks where scores would lie if the result on the first and second test were the same. Points above that line are the scores of children who did better on the second test than on the first, below that line they performed less well on the second test. Presented this way, it's easier to get a quick impression of how many improved and how many got worse. (It also looks like children who did poorly on the first test tended to get better on the second, whereas those that did well tended not to improve as much on the second test.)

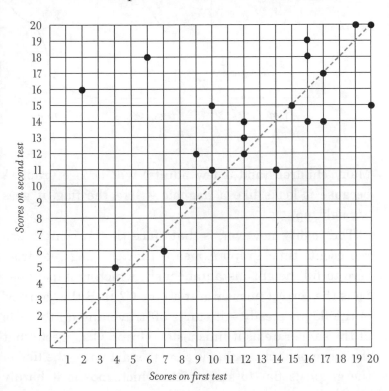

Inside kids' heads

Children can have a problem with interpreting graphs because they think of them as literal pictures of what is going on.

Imagine that a scout is hoisting the Union Jack up a flagpole. Which of these graphs best represents the journey of the flag up the pole? The horizontal axis is time, in seconds, and the vertical axis is the height of the flag in metres.

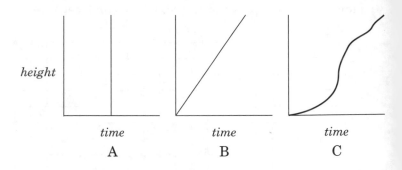

Many children (and a few adults) will choose graph A because it looks like a flagpole – like the flag, it goes 'straight up'.

The reality, however, is that since the horizontal axis represents time, graph A means that the flag goes from zero to full height instantaneously, which is impossible. So is the correct answer graph B or C? In B the 'journey' of the flag is steady – it starts off travelling up the pole at the same rate as it finishes at. Graph C suggests that the movement of the flag is slow to start with (the line of the graph is flat to start with, which means it hardly moves up for the first few seconds), then it speeds up, finally it slows down again. So B might the graph of a flag that is being hoisted mechanically, but C is probably closest to showing what would happen if a human was in control. Making sense of graphs like this takes a lot of work.

Test yourself

ii) *Which sport?*

Which sport do you think this graph represents: golf, 100m sprint or fishing?

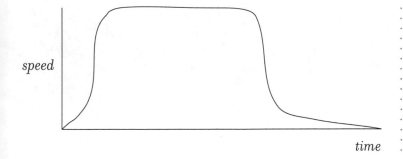

Story graphs

A fun way to explore with your child the fact that graphs are not literal pictures of events is to construct the graph of a character's emotional state during the course of a traditional fairy tale. The X-axis marks the passage of time, while the Y-axis shows emotions from very negative at the bottom, to very positive at the top. Everyone can choose a different story and then try to match the graph to the story. For example, how does this graph relate to little red riding hood's emotions and the events of the story?

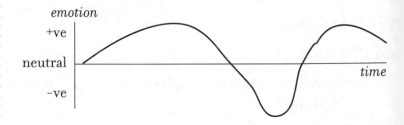

Pie charts

One type of pie chart was invented (or at least popularised) by Florence Nightingale. Proportional data – fractions of a population – are presented in pie charts. For example, the proportions of children travelling to school by different means of transport or the fraction of children having school dinners. Here's a typical sort of pie chart that your child might meet.

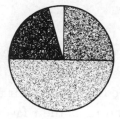 

Questions you might ask:

● Which is the most popular way of travelling to school?
● What percentage of children cycle to school? (the whole 'pie' always represents 100 per cent)

Without further information, however, we cannot say anything about absolute numbers of children. The pie chart informs us that half or 50 per cent of the children walk to school, but we don't know whether that is 20, 100 or 1,000 children, as the total number of children in the data set is not revealed. But if we know that this was a survey of 100 children then other questions can be answered:

● How many children cycle to school? (25)
● 5 children travel by car. How many travel by bus? (20)

Test yourself

iii) *Pie chart*

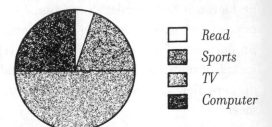

a. What percentage of children chose computer as their favourite pastime?

b. 10 children chose computer as their favourite pastime. 2 children chose reading as their favourite pastime. How many children chose sports?

c. How many children took part in the survey?

Comparing pie charts

Things get more interesting, and more challenging, when presented with two pie charts.

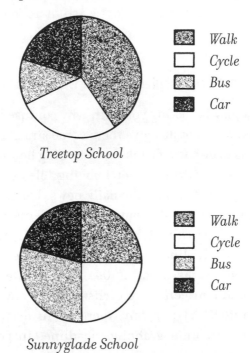

Treetop School

Sunnyglade School

Inspection of these two pie charts reveals that the percentage of children who walk to school is greater at Treetop School than at Sunnyglade. But without further information we cannot say in absolute terms whether or not more children at Treetops walk. If 40 per cent of the children at Treetops walk to school and there are 100 children in the school then that's 40 walkers. While only 25 per cent of the Sunnyglade children walk, if that school has 200 pupils then that's 50 walkers – more children walk to Sunnyglade even though that is a smaller proportion of the overall number of children. In *relative* terms a bigger proportion of the children walk to Treetops but in **absolute** terms more children walk to Sunnyglade. Whenever your child is presented with two pie charts warn them to be on the lookout and to read the question being asked very carefully.

Preparing children for spin doctors

You may be asking yourself, why confuse my child with all this? Why not just give them some straightforward calculations to do? The fact is that learning how to interpret these diagrams is the first step in being able to combat the statistical 'spin' that fills the daily news. For example, the Royal Society for the Prevention of Accidents reported that 136 cyclists were killed in accidents in 2007. This was reported in BBC headlines as reflecting an 11 per cent increase in cyclist deaths since 2004. In absolute numbers that's around fourteen more cyclists. While any deaths are to be regretted, '11 per cent increase in deaths' provides the more attention-grabbing headline. In contrast the BBC reported deaths in the home due to accidents in absolute terms: 76 people *per week*. Compared with an average of less than 3 cycling deaths per week, this absolute figure does sound impressive. Watch out for ministers presenting spending plans in absolute terms but cuts in relative, percentage, terms. Spending an additional £2 million pounds on something sounds impressive, even though it may only be 1 per cent of an overall budget. On the other hand, a cut of 0.5 per cent may appear modest, but could run into millions in absolute terms.

While we aren't suggesting that maths should be used to scare children off their bikes, by the age of nine or ten many children do become interested in topics like climate change or endangered species and they can begin to cope with exploring the different ways that data is presented and the different impressions that can be created by relative or absolute presentations.

MODE, MEDIAN, MEAN AND RANGE

Just about everyone remembers being taught how to calculate the average of a set of numbers: add them all up and divide by the total number of numbers. Strictly speaking this is the 'arithmetic mean' or simply the mean. The mean is the most commonly used type of average and is often referred to simply as the average. Mathematicians and statisticians use the term 'mean' to distinguish it from other averages: the median and the mode. But if most people are happy for the average to mean the 'mean' average, why do kids have to learn about mode and median?

The median value of a data set is simply the middle value: half of the values in a data set are above the median and half are below. Suppose you roll two dice five times, adding the score on each, and get this 'data set':

$$7, 11, 11, 11, 5$$

(Incidentally, this is a real set of data – we did roll the dice and come up with these scores – unusual patterns like this happen more often than people expect them to.)

Putting our dice scores in order:

$$5, 7, 11, 11, 11$$

The **mean** average of these scores is 9 (5 + 7 + 11 + 11 + 11 = 45, 45 ÷ 5 = 9).

The **median** average of these scores is 11 – the middle value. (This is easy to find when there is an odd number of data items. If the dice had been rolled six times, the middle two values are taken – the third and fourth values in that case – and the *mean* of those two values found.)

Finally there is the **mode**. The mode is simply the value that occurs the most frequently in a data set. For our dice scores, the mode was 11, since this score occurred the most often.

Which average is best?

Suppose you work in a men's shoe shop and sell ten pairs of shoes over the course of a morning. The sizes were:

$$8, 7, 9, 6, 9, 8, 10, 8, 10, 6$$

Putting this data set in order:

$$6, 6, 7, 8, 8, 8, 9, 9, 10, 10$$

The mean of this data is 8.1, the median is 8 and the mode is 8. If you need to place an order for more of one size of shoe, all the averages point to ordering some size 8s.

But suppose you sold:

$$4, 5, 5, 6, 6, 7, 8, 8, 8, 8, 8$$

The mean is 7.3, the median 6.5 and the mode 8. Which size shoe should you order in? Extending a data set like this over time, the mode is the most helpful average to use (and yes it's like '*à la mode*' – in fashion, the most popular).

So different averages provide different insights into data and which one you choose really depends on your purpose for finding what is 'typical'.

Another term to know is **range** – that's simply the smallest and largest values of a data set. This can also be helpful in decision making. Suppose in our men's shoe shop, over a six month period, the range of the size of shoes sold is 4–12. Not much point in continuing to buy in men's sizes 3 or 13 is there?

Test yourself

iv) *Dice scores*

What is the mean, median, mode and range of this set of dice scores:

$$4, 11, 8, 6, 5, 6, 9, 11, 7, 2$$

The language of chance

There's a story, probably apocryphal, that a news presenter in the USA was giving the weather report.

'There is a 75 per cent likelihood that it will rain on Saturday. And a 25 per cent chance that it will rain on Sunday. I guess that means there's a 100 per cent chance it will rain sometime over the weekend.'

Unfortunately life, and probability, are not as simple as this. If you knew what the weather was going to be tomorrow, that your mortgage repayments would never change, that your children would be ready for school on time and the car would always start first time. Then again, life would be pretty dull too.

The reality, of course, is that we are surrounded by uncertainty, and being able to cope with events that are not entirely predictable is one of the most important of life skills. That's one reason why, in the enlightened world of modern maths, the study of chance – or, to give it its more formal name, probability – has become an important part of what your children learn.

In primary school, children are only expected to grasp the general principles of probability, not the detailed maths. They come to understand that not all outcomes are certain (for example, it won't necessarily rain tomorrow), and that some outcomes are more likely than others. You can help to reinforce these ideas at home by using the everyday language of probability to describe different outcomes:

- It is *certain* that the sun will rise tomorrow.
- It's *very likely* that Manchester United will do well in the league next season.
- It is *roughly even* (or 'fifty-fifty') that this coin will come up heads when it lands.
- It is *very unlikely* that it will snow in June.
- There *definitely won't* be a gold coin inside this chicken's egg.

Game: Dice bingo

Here's a game of instant bingo that is fun for all the family. The bingo numbers will be decided by rolling two dice and adding up the scores. Draw up one bingo card for each player, with eight squares on each card. Each player can choose which eight numbers to put onto their card, and the numbers should be between 2 and 12 (since the total of two dice is always in this range). When one of your numbers comes up, you cross it off, and the first to cross off all his numbers is the winner. Unlike normal bingo, you are allowed to have a number more than once on your card. You can fill the card entirely with 12s if you wish! But if 12 comes up on the dice, you can only cross off one of your 12s. Your bingo card could look like this:

| 5 | 7 | 8 | 8 |
| 9 | 10 | 12 | 12 |

This game involves a combination of luck and skill. Luck because nobody knows which numbers will come up on the dice, but skill because you can massively increase your chances by picking your numbers carefully.

The diagram opposite shows how many ways there are of scoring between 2 and 12 on two dice. For example, there is only one way of rolling a total of 2 (by rolling two 1s) or 12 (two sixes); there are two ways of rolling 3 (1 and 2 or 2 and 1). On the other hand there are six ways of rolling a total of 7: 1 and 6, 2 and 5, 3 and 4, 4 and 3, 5 and 2, 6 and 1. So you are six times as likely to roll a total of seven as a total of two.

So choosing to fill your card with 12s would not be very sensible – your chance of them all coming up within a few throws is minimal. But should you fill the card with sevens? Well, that mightn't be so clever either. Just because 7s are more likely, doesn't mean that they will appear every time. Your best bet is to choose a combination of 6s, 7s and 8s (with maybe a 5 and a 9 thrown in).

Number of ways of getting score

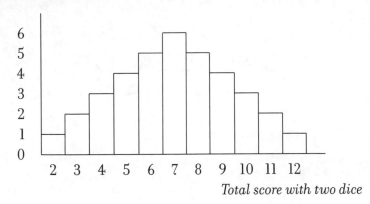

Total score with two dice

CALCULATOR MATHS

Q. Use a rule to measure this line:

Be kind to pepol

This child thought of a good rule – 'be kind to people' – and made sure it fitted the line exactly so that (in her opinion) it was measuring the line.

I t wasn't until the late 1970s that calculators first became cheap enough that every household might expect to own one. That means that many parents – and almost all grandparents – had little, if any, experience of calculators at school. Today, of course, calculators are a standard piece of equipment in the classroom.

In this chapter we look at the traps that lurk within even the simplest calculator, and also at how calculators can help, as well as hinder, your child's mathematical understanding.

Common problems children have with calculators

1. Not questioning the answer that the calculator gives, no matter how spurious.
2. The order (or method) of doing calculations may be different from the way they do it manually.
3. Being worried by strange symbols on some of the keys.
4. Forgetting to clear a previous calculation before doing the next one.
5. Clumsy fingers pressing the wrong keys.
6. Forgetting to press the decimal point button (because a decimal point already appears at the end of the number on the display).

Do calculators damage your child's maths?

It's easy to put the nation's supposed decline in mathematical ability down to calculators, but it's not as simple as that. For a start, there are many aspects of maths in which calculators are of no use whatsoever. Just take a look at the questions at the end of the book to see how often a calculator is little help in finding a solution. (Academic mathematicians often claim they don't even own a calculator, so rarely do they have call for any 'calculations' in their work.)

But calculators *can* be a problem. The very fact that they exist means that there is little appetite in schools for children to practise hundreds and hundreds of sums like their [grand]parents did, because what's the point if they'll rarely need to do these sums as adults? This means that techniques and number patterns that were embedded in the minds of previous generations tend not to be embedded in children's minds today.

The other problem is that children – and sometimes their parents – do have a tendency to believe answers that their calculators give them, because of their belief that calculators never go wrong. And this is true, up to a point, because the technology is almost faultless. But a calculator has no idea if the person using it has put the decimal point in the right place or pressed the right numbers, let alone whether the person using it has correctly interpreted the problem they are trying to solve in the first place. It brings to mind the old expression made popular in the early days of computers: 'Garbage in, garbage out.'

Calculators are usually first introduced in the classroom when children are about eight years old. By then, the children ought to be familiar with numbers, place value, addition and their tables, so the main functions of a simple calculator should be meaningful. But the chances are that children will encounter calculators before that, so a bit of indoctrination in the good and bad aspects of calculators can be useful from an early age.

Is the calculator right?

You can begin to sow the seeds of healthy mistrust in the calculator at an early age, by playing the 'Is the calculator right?' game. You can play it as soon as your child is able to add up. 'What's four plus two?' you ask. 'Let's see what the calculator says . . . it says six. Is that right? I'm not sure, let's check . . .' When you agree that six is the answer, grudgingly accept that the calculator has got it right this time, but that you will be keeping your eye on it in case it gets any funny ideas . . .

Even better, before introducing your children to the electronic version of a calculator, you can create a homemade version . . .

Game: Matchbox calculator

This trick is a real groaner, but it's a great way to get them to practise doing calculations! You will need an ordinary, empty matchbox. On the back of the box, draw the calculator numbers 0 to 9 and the main operations as if it is a genuine calculator. Show the matchbox to your child, and say: 'This may look like an ordinary matchbox, but would you believe that it can do amazing calculations.'

Think of a sum that she ought to know the answer to, and pretend to put this sum into your matchbox as if you were pressing the keys of a calculator. Say the sum out loud as you press the 'keys'. For example, 'Seven plus five equals . . .' Ask your child the answer. If she says 12, push up the tray to reveal the word you secretly wrote there beforehand: 'CORRECT'. If she gets the wrong answer, push the tray the other way to reveal the word 'WRONG'. You can heighten their curiosity by keeping the 'display' hidden for the first few calculations.

The pitfalls of the simple calculator

The calculator that your child is first exposed to at school will be something like the one opposite. It's so basic, what can possibly go wrong when your child first has a go? Actually, a lot.

For a start, there is a remarkable amount of variety even in basic calculators, not only in the positions of the various buttons but also in the keys on offer. Some calculators don't have an OFF button, others have an OFF but not an ON, some have a special $+/-$ key in addition to the $+$ and $-$ keys, and there can be a whole variety of keys with 'M' on them, including one of $\boxed{\text{RCM}}$, $\boxed{\text{MRC}}$ and $\boxed{\text{MR}}$, each of which means 'Memory Recall'. And just when you think a calculator can contain no more surprises, you'll find one with an $\boxed{\text{MU}}$ key, which is nothing to do with Memory but actually means 'Mark Up' – which might be handy for a few retailers but is a complete black box to anyone else.

Display. Usually 8 digits. Basic calculators display an error message (often in small writing, so children miss it) when a calculation gives an answer bigger than 99999999.

The percentage key – should be banned from calculators (see below).

+/− key. This changes a number from positive to negative, and is usually put well away from the addition and subtraction keys to avoid confusion, but some children are still caught out.

Decimal point. Many calculators automatically show a decimal point at the end of the number on the display, which means children sometimes assume they won't need to press the [point] button to write a decimal.

Square root. Many children leave primary school not knowing what this button is. Basic calculators do square roots 'the wrong way round'. When working out a square root you say 'the square root of 25 equals . . .' (answer 5), but on a calculator you have to work in reverse, entering , '25 √'.

Clear All. It should become second nature for children to press this button before they start a calculation. Otherwise it's possible that the calculator with 0 on the display is actually in the middle of a previous calculation, in which case a spurious answer will appear.

Equals. On a calculator, this button means 'give me the answer' rather than its true meaning 'is the same as'. This can lead to bad habits that affect the way children understand maths problems such as '2 + K = 20'.

So with calculators it's certainly not the case that when you've seen one, you've seen them all. And while children have a reputation for being able to figure out electronic gadgets far faster than their parents, there are lots of pitfalls to be aware of. The diagram on the previous page highlights some of the most common ones to be found in basic, cheap calculators.

The percentage button

Most calculators have on them an innocent-looking button with a percentage label. But the truth is that you should use this button at your peril. Indeed, there are some maths teachers who treat the percentage button like the Forbidden Fruit of the Garden of Eden and tell their children: 'You can use any button you like except for one – NEVER press per cent.' Why is this? Because you will only get the correct answer if you use the button the right way, and it is easy to get it wrong.

For example, what if you want to work out four fifths as a percentage? If you make the mistake of pressing the equals button (which children usually expect to press to get an answer) then on most calculators, here are the answers you get:

BUTTONS BEING PRESSED:	CALCULATOR ANSWER:
% 4 ÷ 5 =	0.8
4 % ÷ 5 =	0.8
4 ÷ % 5 =	1.25
4 ÷ 5 % =	16

On most calculators, the correct way of finding $4 \div 5$ as a percentage is to press $\boxed{4}\boxed{\div}\boxed{5}\boxed{\%}$ and not press the equals button (you can confirm this on yours – the answer 80 should comes up at the end, if it doesn't then your calculator doesn't have the standard logic). But the fact that you have to press different buttons in a different order on some calculators is warning enough that the % button should be avoided at all costs.

If you do want to use a calculator to solve a percentage question, then this is how most basic calculators are used to solve the most common problems:

Calculate three sixteenths as a percentage:

$\boxed{3}\boxed{\div}\boxed{1}\boxed{6}\boxed{\%}$ Result: 18.75%

Calculate 17.5% of £250:

$\boxed{2}\boxed{5}\boxed{0}\boxed{\times}\boxed{1}\boxed{7}\boxed{.}\boxed{5}\boxed{\%}$ Result: £43.75

Add 15% VAT to £20:

$\boxed{2}\boxed{0}\boxed{+}\boxed{1}\boxed{5}\boxed{\%}$ Result: £23

Give 10% discount on goods costing £100:

$\boxed{1}\boxed{0}\boxed{0}\boxed{-}\boxed{1}\boxed{0}\boxed{\%}$ Result: £90

Three ways to show up a calculator

If you need more evidence of the potential pitfalls of a calculator, try out these little exercises.

1. Enter the calculation $\boxed{1} \boxed{\div} \boxed{9}$. Multiply the answer by 9. Finally subtract 1 from the answer. Everyone knows that $1 \div 9 \times 9 = 1$, and that $1 - 1 = 0$, but if you have a regular (cheap) calculator it is convinced that in this case $1 - 1 = -0.00000001$.

2. A more extreme test of your calculator's hidden quality (or lack of) is to divide 1 by 11, then multiply the answer by 8 and then by 11. The answer should be 8 (the 11s cancel, making the sum 1×8), but typically a calculator does no better than 7.999992 — an example of a simple number being brutally massacred by electronics.

3. Ask your calculator to tell you if 87654321×12345678 is even. It doesn't know — it can probably round the number up, but can't tell you the final digit. Whereas of course you know that any number multiplied by an even number must also be even (and you can even be confident, using the ideas of long multiplication, that the final digit of this sum is going to be 8).

BODMAS

Calculators have in-built rules for the order in which they do calculations. Unfortunately, in most basic calculators this order is different from the order that your children will be taught at school.

What is $3 + 4 \times 5$? Is it $3 + 4$ (i.e. 7) $\times 5$, which is 35? Or is it $3 + (4 \times 5)$, which is 23? Check which answer your calculator gives – most likely, it will give the answer 35, because it carries out each instruction in turn. Unfortunately this isn't the answer that will be expected from your children.

It's clearly very important that everyone should use the same rule, or we'd all come up with different answers. For this reason, there is a universal convention (followed by everyone except basic calculators) that when doing a calculation with more than one operation, you should do things in a particular order:

- First do any calculations that are inside brackets. So if the sum is $3 + 4 \times (6-1)$, the first thing you do is $(6-1)$, which is 5.
- Next you do all the multiplications and divisions (do these in the order that they appear, so for example to calculate $12 \div 3 \times 4$, first work out $12 \div 3$ $(= 4)$ then multiply by 4 (answer 16).
- Finally you do the additions and subtractions (like multiplication and division, do them in the order they appear).

So when calculating $3 + 4 \times 5$, you should do the multiplication first $(4 \times 5 = 20)$ and then add the 3, to give 23.

There is a little aide-memoire that you might have encountered at school that is meant to make this rule stick. It is often called BODMAS, though some people call it BIDMAS and even BEDMAS. Whatever you call it, the crucial thing is that B(rackets) come first and A(ddition) and S(ubtraction) come last.

In case you are wondering why people can't agree on the second letter, it's because, like many acronyms, this one is a bit of a fudge. The E in BEDMAS refers to the numbers that have 'exponents'. For example 5×5 is usually written as 5^2 (or 5 squared), and the small 2 is the exponent. In the world of mathematics calculations, Exponents come after Brackets but before Division and Multiplication.

So for example $3 + 4 \times (2 + 3)^2$ would be calculated as:

$2 + 3 = 5$	(B comes first)
Then $5^2 = 25$	(E comes second)
$4 \times 25 = 100$	(M comes third)
$3 + 100 = 103$	(A comes last)

People sometimes call the exponent of a number its 'Index', others may call it the 'Order', hence the alternatives of BIDMAS and BODMAS. Some people even ignore the exponents altogether and read BODMAS as Brackets Over Division, Multiplication, Addition, Subtraction.

Test yourself

i) *BODMAS*

a. $14 + (7 - 6) \times 2$ b. $3 \times 71 - (208 \div 4)$

c. $2 \times 6 + (83 + 7) \div 9$

Game: Beat the calculator

You can demonstrate that sometimes the mind can work faster than the calculator with this neat addition trick, in which you appear to add up five huge numbers almost instantaneously.

● Get your child to write down any two four-figure numbers. For example:

Child's first number: 5 0 3 8
Child's second number: 6 6 3 5

● You then write down one number under his:

Your first number: 4 9 6 1

(Unknown to him, you make sure that this number added to his first number makes 9,999. This is easy to do — take his first digit (5) and make your first digit 4 so that they add up to 9, and do this for each other digit.)

● Ask him to write down one more number. He might write:

Child's third number: 8 9 2 4

● Now you write a second number.

Your second number: 3 3 6 4

(This number added to your child's second number makes 9,999.)

- And now you issue the challenge, you bet that you can add up all five numbers in your head faster than he can do it on a calculator. While he scrambles for the buttons, you calmly mutter a couple of numbers to yourself and write out the answer:

$$28922$$

To get this number, put 2 in front of your child's third number (in this case it was 8924), and take two off the last digit of his number.

How does it work? All you have done is add 9999 + 9999 to his third number, which is the same as adding 20,000 − 2 + his third number.

This trick demonstrates how maths isn't just about number-crunching, it's also about thinking *smart*.

Discovering number patterns

So far, calculators have taken a bit of a bashing in this chapter. But it is certainly not all bad news. In fact, we are fans of calculators. Used properly, calculators have taken unnecessary drudgery out of mathematics. They can also bring numbers to life for children of all abilities. They can be a fast route to discovering some enchanting number patterns that, without calculators, would have been a closed book to children in past generations.

For example, pick a number between 1 and 9. Now multiply it by 37. Then multiply it by 3. Your original number will appear three times on the screen. So if you pick the number four, then $4 \times 37 \times 3 = 444$. Children like discovering patterns like this, and are likely to want to repeat the exercise themselves.

Now choose any number between 1 and 10. Multiply by the numbers 3, 7, 11, 13 and 37 *in any order*. The number you first thought of appears six times. For example if you pick 5, then . . .

$$5 \times 3 \times 11 \times 13 \times 37 \times 7 = 555555$$

The mysterious 12345679

12345679 is a curious number. Look what happens when you multiply it by numbers from 1 to 9:

$12345679 \times 2 = 24691358$ (contains all digits except for 7)

$12345679 \times 3 = 37037037$

$12345679 \times 4 = 49382716$ (contains all digits except for 5)

$12345679 \times 5 = 61728395$ (the reverse of the one above, except 4 turns into 5!)

$12345679 \times 6 = 74074074$

$12345679 \times 7 = 86419753$ (contains every digit except for 2)

$12345679 \times 8 = 98765432$ (contains all the digits except for 1)

And finally:

$12345679 \times 9 = 111\,111\,111$ (wow!)

Helping with tables

A calculator can be used to help accompany the repetition of times tables (though this does only work on very basic calculators). To do the 3 times table, for example, simply press these buttons in order:

$\boxed{3}\boxed{+}\boxed{=}$ (You may need to press $\boxed{=}$ twice on some calculators.)

This gives the answer 3.

Now press the $\boxed{=}$ again and you get 6. If you now keep pressing $\boxed{=}$, you work your way up the table 3, 6, 9, 12, 15 . . . To get other tables, simply replace 3 with the number whose table you want. Once you've set up the first number, you can leave your child to press $\boxed{=}$, and the answer on the display helps to reinforce his recall of the numbers.

Game: Broken six

In this game you pretend that the six button on the calculator is broken — six can appear on the screen, but (in the game) hitting the six button will not bring it up. This game works at any level. How would you use your broken calculator to calculate things like 11 take away 6? (There are lots of ways of course, including 12 take away 7, for example). You can make it really tricky. For example, what is 676 divided by 16? Whatever solution you find, you're going to need some mental arithmetic to get there. In this case, 676 divided by 16 is the same as, for example, 338 divided by 8 (both numbers have been halved).

Getting creative with your calculator – turning sums into messages

Some calculator numbers become letters if you turn them upside down – maybe you remember as a child discovering you could get the words ShELL.OIL to appear on the inverted display. This still works on most basic calculators, because their numbers are made up from a grid:

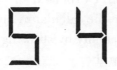

So the number 5 resembles an 'S' and the number 4 upside down is like an 'h'. (On more sophisticated modern calculators, the number four appears like this 4 – which doesn't resemble an h when upside down – while the upside down curvy 5 barely resembles an 'S'.)

You can make the inverted letters game into a challenge that will generate some laughs and get children experimenting. Show them these four riddles, which they have to solve by calculating the answers:

1. The more you take away, the bigger it gets. What is it? 463×8
2. If Sally eats five sausages and four bars of chocolate and then drinks eight cans of lemonade, what does she become? 257×3
3. What was the first thing that the British Prime Minister said when he met the US President? $1{,}289 \times 6 \div 10{,}000$
4. And what did the President say back? $5 \times 802 - 4{,}000 + 4$

Having given them the idea, you can now ask them to come up with their own mystery riddles and sums for *you* to work out, a painless way of getting them to experiment with all sorts of complicated calculations.

Investigating the square root button

Primary children aren't expected to be familiar with the square root button on the calculator, but it does no harm at all to arouse their curiosity by showing them that it can do strange things to a number.

The first thing they can discover is that if they enter a big number and then press the root button, the number changes – and they haven't even pressed equals. So, for example, enter 100 $\boxed{\sqrt{}}$ and the calculator displays 10. You can also point out that as long as they started with a number bigger than one, after pressing the $\boxed{\sqrt{}}$ button the number got smaller, though for most numbers picked at random, there will be lots of digits after the decimal point.

For square numbers such as 9, 25 and 144, pressing the square root produces a pleasing whole number, and they ought to quickly pick up the idea that the square root button is just the reverse of squaring: $\sqrt{}$ 9 is 3, and $3 \times 3 = 9$.

Estimating the square root

Before introducing the square root button an interesting challenge is to explore square roots with your child through approximations. Secretly put a number into the calculator, say, 15, and multiply it by itself. Hand the calculator to your child – can they figure out what your original number was? They will do this through trial and improvement – trying a number, squaring it and then adjusting the trial number in the light of whether the answer was too big or too small. Exploring in this way what number multiplied by itself gives 10 as the answer will get your children exploring the nature of numbers with an increasing number of decimal places.

Another similar investigation is to secretly multiply together two consecutive numbers, say 36 and 37. Handed the calculator showing 1,332 can your child figure out the two numbers?

Test yourself

ii) *Consecutive numbers*

What two consecutive numbers multiplied together give a product of 4,692?

Game: Square roots and candles

● Enter the number 390625 onto a calculator. To make this number more interesting, you can tell a story around it. Start by saying that you have ten invisible boxes, numbered 0 to 9, and that each box contains a different coloured candle. Ask your child to pick a colour. 'Pink.' 'Aha, that's in box 3,' you say – in fact you say that whatever colour they nominate. (Enter 3 on the calculator). 'Another colour?' 'Blue.' 'That's in box 9, let's put that in.' (Enter 9). As your child thinks of a different colour candle, they of course happen to be in boxes 3, 9, 0, 6, 2 and 5, in that order.

● Now with your thumb secretly poised over the square root button, ask your child to blow out the candles. As they blow the first time, secretly press 'square root'. The 3, 9 and 0 will be 'extinguished', leaving just 625. Ask them to blow again. This time the 6 is extinguished. Then again, and the 2 is blown out. Ask them for one finally big puff, and as they do it, move your thumb to the OFF switch and the screen will clear. (Calculators with solar panels sometimes don't have OFF switches, in which case the only way to completely blank the calculator display is to blackout the room!)

The reason why this works is that 5, 25 and 625 have the unusual property that when you square them, the original number appears at the end of the answer.

Game: Six-One-Six

This doesn't need to be a calculator game, but you can make use of a sophisticated calculator (one that allows you to raise numbers to exponential powers) for the later stages.

Prepare three cards, with the numbers 1, 6 and 6 on them (writing the '1' as a simple vertical line). Challenge your child to make these three cards into the biggest number possible. Then, when they've done that, challenge them to make the smallest number possible.

They might well come up with 661 as the biggest number and 166 as the smallest. Those are certainly good answers. But there's lots of scope to be creative here. What about turning the 6 upside down? Now it becomes a 9, so you can make 991 as the biggest number.

And it goes further. What if you turn the 1 on its side? Now you can have $9/9$ which is 1. Or if you treat the 1 as a minus sign, you have $9 - 9 = 0$. Even 'smaller' than zero (if you think of negative numbers as being smaller than zero) you can go to -99. And if you really want to stimulate your child's thinking, and you think they are ready for it, you can even introduce the idea of powers. 9^2 means 9×9, and in the same way, 9^6 means $9 \times 9 \times 9 \times 9 \times 9 \times 9$. And that means if you want to go really big, you can write 9^{91}, which represents a number so big that it is bigger than the number of atoms in the universe. Meanwhile, the lowest you can go with the symbols is -9^9, which works out at minus 387 million. That's the sort of number that makes banks very nervous.

There's more about powers in the next chapter.

BIG IDEAS
FOR SMALL PEOPLE

Q. Draw a shape with three lines of symmetry

This child confused symmetry with cemetery.

Most of this book is devoted to the maths that your child will first encounter at primary school. But what about the really meaty stuff – algebra, geometry, logarithms and the notion of infinity, for example? That's for big school, isn't it? That's certainly where the curriculum puts it, but there are many ideas from 'higher maths' that a bright ten-year-old can not only understand, but can even get excited about. In this chapter we'll give you seven ways to introduce some higher maths to your child, including a little bit of 'magic' . . .

Think of a number (the magic of algebra)

Think of a number, any number you like (but you might want to make it less than ten).

Double it.

Add 10.

Divide your answer by two.

Finally take away the number you first thought of.

And we predict that the number you finished at was . . . five.

It certainly should have been.

Children love this 'trick' because it feels like mind-reading. (But be careful, children younger than nine are quite likely to make a mistake in their calculations, and so end up with a number different from five, which does rather remove the magic.) They like to test it again and again to see if they can 'catch it out'. As they get more adventurous, they might think they have found a number that defeats the trick, but this is always down to an arithmetical mistake (so if they don't end up on five, get them to tell you which number they started with, work through the calculations with them, and then with a smile 'discover' that their number really did end up at five after all). And it really does always finish at five, even if you start with a big number, a decimal, or a negative number.

Doing the calculations is good practice for basic mental numeracy skills, but perhaps your child is curious to know *why* it works. You can explain it by giving your child's number a name. Why not call it something that your child will find funny, like 'Blob', and imagine that the number Blob is sealed tightly in an envelope. We'll now do the trick using Blob.

- Think of a number: *Blob*.
- Double it. What do you get? Blob Blob (or *two Blobs*), which means there are now two envelopes with Blob inside.
- Add ten – these can be ten of anything, but let's think of them as ten fingers. You now have *two Blobs plus ten* fingers.
- Divide your answer by two. Half of 'Two Blobs and ten fingers' is 'One Blob and 5 fingers'.
- Finally take away the number you first thought of, which was Blob. 'One Blob and five fingers', take away Blob, leaves you with five fingers.

In other words, in this little trick, it doesn't matter what number you start with ('Blob') because you always get rid of Blob in the end, leaving you with the number 5 (fingers).

We've used Blob to represent any number, and what we're doing here usually goes by the formal name of algebra. The main difference between this algebra and the algebra taught in secondary school is that in school Blob is given a less amusing label such as x or y. There's a common complaint among adults about how, at school, they never understood why numbers had to be replaced by letters. The Blob magic trick is one way to demonstrate how giving unknown numbers a name can be useful.

Incidentally, once you've seen how this basic version of the trick works, you can explore what happens when you change the instructions. For example, how can you change the original trick so that it always ends up with the number 6? (Answer: add 12 instead of 10.) What if you treble instead of double? The potential is endless.

Colouring pictures

Maths can crop up in the most unlikely places. One example is colouring maps. When you colour a map, you don't want two adjacent regions to be the same colour. It took mathematicians over 100 years to prove that you only ever need, at most, four different colours to fill a map so that no bordering regions are the same colour (regions that only touch at a point are allowed to be the same). Check this for yourself: you can shade Australia's regions entirely using only red, green, yellow and blue (for the sea).

There's a variation on this that is less well known. On a blank piece of paper, do a scribble, keeping your pen on the paper at all times, and ending the doodle at the same point as you start. You'll end up with something like this:

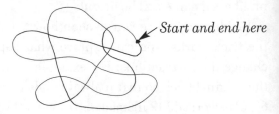

Start and end here

How many colours do you need for this map? You will only ever need two — there is no scribble that ever needs more. For example you can make your scribble black and white with no neighbours the same colour:

Any child can investigate it. But proving why this is so requires some sophisticated maths known as graph theory — fascinating, but way beyond primary maths.

Mind-reading cards (the magic of Binary numbers)

They say that the old ones are the best, and that is certainly the case for this popular 'mind-reading' trick that is often to be found in Christmas crackers or children's magic sets. You need to have four cards with numbers on, exactly as shown here.

First card

8	9	10	11
12	13	14	15

Second card

4	5	6	7
12	13	14	15

Third card

2	3	6	7
10	11	14	15

Fourth card

1	3	5	7
9	11	13	15

The script goes like this: 'Think of a number between 1 and 15, but don't tell me what it is. Now I show you each of your four cards in turn, and for each one I ask you, "Is your number on this card?" If you say "yes" for any card, I keep that card to one side. When you have shown me each of the four cards, I can now magically tell you what your number is!'

To find your number, take the cards that have your number on and add together the numbers at the *top left corner* of those cards.

Here's an example. Suppose you think of 13. This number appears on all of the cards except for the third one. The corner numbers of the cards that 13 is on are: 1, 4 and 8. Add these together to get: 1 + 4 + 8 = 13. So it works! But *why*?

Take a look at this series of numbers:

$$1, 2, 4, 8, 16, 32, 64$$

These are called powers of two. Each number is double the one before, and these numbers are the secret behind the mind-reading card trick. The first number on each of the cards is a power of two.

It turns out that you can make up any whole number by adding different powers of two together (see 'Grandma's weights' on page 241). For example:

To make 6, add 4 plus 2
To make 9, add 8 plus 1
To make 14, add 8 plus 4 plus 2

This is how to make all the numbers from 1 to 15:

	8	4	2	1
1	No	No	No	Yes
2	No	No	Yes	No
3	No	No	Yes	Yes
4	No	Yes	No	No
5	No	Yes	No	Yes
6	No	Yes	Yes	No
7	No	Yes	Yes	Yes
8	Yes	No	No	No
9	Yes	No	No	Yes
10	Yes	No	Yes	No
11	Yes	No	Yes	Yes
12	Yes	Yes	No	No
13	Yes	Yes	No	Yes
14	Yes	Yes	Yes	No
15	Yes	Yes	Yes	Yes

Each number can be uniquely described by a combination of Yes's and No's. For example, the number 3 is 'No No Yes Yes' while 13 is 'Yes Yes No Yes'. Replace Yes with a 1 and No with a 0 and you get 0011 for 3 and 1101 for 13. These are the so-called binary numbers that are fundamental numbers used by computers, because computers at their most basic level work entirely on making yes/no decisions. That makes binary numbers extremely important, perhaps the most important number system on the planet.

To decide which numbers should go on which card, simply look for the Yeses in each column in the table. The first column (8) has a Yes against numbers 8, 9, 10, 11, 12, 13,

14 and 15, so these are the numbers that go on the first card. The second column (4) has a Yes against 4, 5, 6, 7, 12, 13, 14, 15 and these numbers go on the second card. And so on.

You can make this trick even more impressive by using five cards instead of four. To work out which numbers go on which card, you need to draw up a bigger table. It needs one more column (with 16 at the top) and the numbers down the side of the table should go from 1 to 31 (31 is the largest number you can get by adding 1, 2, 4, 8 and 16). The five cards that you end up with are:

```
 1  3  5  7  9 11     2  3  6  7 10 11     4  5  6  7 12 13
13 15 17 19 21 23    14 15 18 19 22       14 15 20 21 22
25 27 29 31          23 26 27 30 31       23 28 29 30 31
```

```
 8  9 10 11 12 13    16 17 18 19 20 21
14 15 24 25 26       22 23 24 25 26 27
27 28 29 30 31       28 29 30 31
```

Now do the trick, getting your 'victim' to choose a number between 1 and 31.

The power of powers

How many grains of sand would it take to fill the whole observable universe? It's a silly question, since there isn't enough sand available to fill the universe (where would the sand come from?) but children love silly questions. And this one has a silly sounding answer too: it would take about a million billion billion trillion quadrillion pentillion hextillion grains of sand . . . times three! Or if we write it down, that is 3 followed by 90 zeroes, which looks like this:

3,000,000,000,000,000,000,000,000,000,000,000,
000,000,000,000,000,000,000,000,000,000,000,
000,000,000,000,000,000,000,000

That's a long number to write out, but fortunately mathe maticians have a shorthand code. This gigantic number can be written as 3×10^{90}. The little index number 90 is known as the power, or exponent, and is also sometimes known as (watch out, scary word coming up) the *logarithm* (in base 10).

Most parents will remember encountering logarithms at school, but will be sketchy at best about what they are and how they work. But the idea behind them can be introduced to a ten-year-old child.

Start by reminding them that the area of a square whose sides are 10 metres is 10×10 , or '10 squared'. In shorthand this is written as 10^2 (which makes sense, because the number 10 appears twice in the sum).

So it also makes sense that $10 \times 10 \times 10$ is written as 10^3.

Now what is $10^3 \times 10^2$? Written out longhand it is $10 \times 10 \times 10 \times 10 \times 10$, or 100,000, but in shorthand this is 10^5. Notice how the little numbers in the original multiplication (3 and 2) have simply been added together to give 5 in the answer?

Does this adding idea always work? What might you expect $10^2 \times 10^4$ to be? If the adding rule works, then the answer should be 10^6 because $2 + 4 = 6$ – and a quick check should confirm that this is the right answer ($100 \times 10,000 = 1,000,000$, or a million).

So logarithms can be used to turn multiplication into addition. And they work for any base number. So $3^2 \times 3^4 =$

3^6 $(2 + 4 = 6)$. Written longhand, that's 3×3 (9) multiplied by $3 \times 3 \times 3 \times 3$ (81) equals $3 \times 3 \times 3 \times 3 \times 3 \times 3$ (729). This becomes useful when the numbers get too big for a calculator. So you can now confidently say that $17^9 \times 17^4$ equals 17^{13} even though this sum probably produces the message 'error' if you try to check it electronically.

If your child is prepared to accept that this adding rule is a good thing, and that it always seems to work, you can use it to extend the idea of powers even further. What does 3^0 mean? Most children (and most adults come to that) might expect it to be zero, but if you accept that the adding rule works, then 3^0 must equal 1. Why is this? Think about the calculation $3^0 \times 3^2$. Because of the adding rule, $0 + 2 = 2$ so the answer must be 3^2, in other words $3^0 \times 9 = 9 \ldots$ which means 3^0 must equal 1. In fact, according to this rule, *any number raised to the power of zero* is equal to 1!

This may require a bit of a breather to take in, and you might want to stop at this point. But just in case you or your child are clamouring for more, you might like to know that the adding rule works for fractions and negative numbers too. For example, $10^{1/2}$ is the square root of 10 (about 3.16) because $10^{1/2} \times 10^{1/2} = 10^1$. And how about negative numbers? 10^{-1} is 0.1, or one tenth, because $10^{-1} \times 10^1 = 10^0 = 1$. OK, enough!!!

These are big ideas – maybe some of them are too big for many children. But don't forget that most children are fascinated by huge numbers, and the fact that numbers the size of the universe can be written in just three or four digits (such as 10^{91}) is a mathematical mystery worth sharing.

Finding the area of a triangle

Let's get back to everyday numbers again. On page 237, we described the way of approximately working out the area of a triangle by counting squares. But, as many parents will be aware, there is a rule for calculating the *precise* area of a triangle, which you might remember as 'half base × height'.

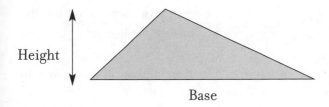

Height

Base

So if the base of the triangle here is 6 metres and the height is 3 metres, the area of the triangle is $\frac{1}{2} \times 3 \times 6 = 9$ square metres.

But *why* is the area of a triangle always half base times height? Few parents recall having this explained, they just remember the rule.

In fact, the explanation is very simple, and once you understand it, you can enlighten your children. Take any triangle you like, and imagine laying it down on its longest side:

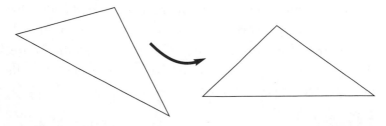

Now imagine placing your triangle inside a tightly fitting box and draw a vertical line from the top (apex) of the triangle to the base:

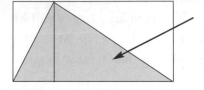

Each part of the triangle is half of a rectangle, so the total area will be half of the area of the whole rectangle – in other words, half of its base × its height.

Placing these rectangles together helps to explain why triangles always tessellate.

Place two of the original triangles side by side. The gap between the two triangles will always be an identical copy of the original triangle flipped upside down, so the triangles will fit into a rectangular strip that can continue for ever.

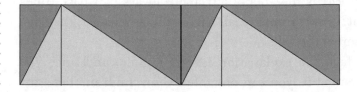

This is an example of a visual proof – and proof is one of the most important principles in mathematics, the idea that you can say, with certainty, that for any shape of triangle anywhere in the world, today, in history and in the future, the area has and will always be half of its base times its height. That's a powerful idea.

What makes circles so special?

There is an old puzzle in which Farmer Giles has a goat and 120 metres of fencing. The farmer wants to enclose a field with as much grass as possible.

First he tries an equilateral triangle, with each side 40 metres (so the perimeter is 120 metres):

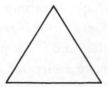

The area is just under 700 square metres (the base is 40 metres and the height turns out to be roughly 35).

Can he beat that? He arranges the 120 metres of fencing into a square, 30 × 30 metres:

Now the area is 30 × 30 = 900 m². So if you have a certain length of fence, square fields enclose more area than triangles.

Next he tries a regular hexagon, with each side 20 metres:

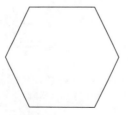

Now the area is about 1,039 square metres (one way to calculate this area is to work out the area of each equilateral triangle segment inside the hexagon, and multiply by six).

So for a given amount of fence, it seems that you enclose more area if you keep adding more sides to the field.

Take this further and pass through the decagon (10 sides), and the icosagon (20 sides) and as the number of sides increases, the shape of Farmer Giles' field begins to resemble a circle. In fact a circle is really a polygon with an *infinite* number of sides, and if Farmer Giles uses his 120 metres of fence to make a circular field, he can enclose nearly 1,150 square metres, far more than his original triangle, and in fact it's the maximum possible area that can be enclosed. That's just one of the many important properties of circles, and why they feature so prominently in geometry.

Incidentally, this notion of working out an answer by seeing what happens when you take smaller and smaller steps until you get to the infinitely small is the basis of one of the most important areas of higher mathematics, known as calculus. Much of our understanding of higher maths owes everything to calculus, and we have Isaac Newton to thank (among others) for his ideas about the maths of the infinitely small. Which leads to one final topic . . .

Infinity and beyond

Next time your children are watching *Toy Story*, or playing with their Buzz Lightyear action figure, ask them about his catchphrase: 'To infinity and beyond.' Children get excited by the idea of infinity from as young as five or six years old, because it's the 'biggest' number possible. Or is it? Buzz Lightyear seems to be claiming that you can go beyond infinity.

Infinity is certainly a strange idea, and you can get an insight into this by telling the story of Hilbert's Hotel.

There was once a hotel (called Hilbert's Hotel) that had infinite rooms. Incredibly, one night the hotel was full. There was somebody in room 1, 2, 3, 4 all the way up to infinity. Then a man turned up at the entrance and asked, 'Do you have a room?' The manager thought for a moment, and said, 'Luckily for you, yes I do.' The manager sent a message to all his customers: 'Please move to the room number that is one higher than the room you are in now.' The person in Room 1 moved to Room 2, Room 2 moved to 3, 3 moved to 4, and so on. And of course for any number you think of, however big, there is always a number one higher. So everyone now had a room, and every room was full – except for Room 1 that was now empty. The manager handed the key to Room 1 to the extra customer.

In the weird maths of infinity, you now have an answer to the question, 'What is infinity plus 1?': Infinity plus 1 *equals* infinity. Most secondary pupils, even sixth-formers, never come across the idea of infinity plus one and Hilbert's Hotel. These ideas are typically first discussed at university. But you'll find that an eight-year-old will engage with it too.

You can take the story further. The next day, a bus turned up, and this time there were an infinite number of people in the bus. The hotel was still full, so what could they do? Fortunately the manager had another plan. This time he sent a message to all his customers asking them to move to the room that was twice the number of the room they were in at the moment. Room 1 went to Room 2, Room 2 went to 4, and so on. And since every number always has another number that is double it, everyone found a room. If you draw two number lines, you can see what happened:

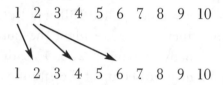

All the customers in the hotel moved to an even-numbered room. That means that all the odd-numbered rooms were now empty. And since there is an infinite number of odd numbers, everyone on the infinite coach was able to take a room. That means two times infinity is also infinity.

It really does seem as if it isn't possible to go beyond infinity. Well, as it happens, that's not true. If a bus turns up containing infinite passengers who have T-shirts with different numbers on, and those numbers are every single possible decimal number between zero and one, then this time there are more people than there are hotel rooms. There really are infinities that are bigger than the one we think of when we are counting. But that's such a big idea, maybe it's one that should wait until your children are a little bit older.

THE
QUESTIONS

QUESTIONS THAT YOUR 11-YEAR-OLD MIGHT ENCOUNTER

At the age of eleven (give or take a year), your child is going to be tested. This might be the government's national tests, or it might be an entrance exam for a selective school. Whatever the tests or exams that your child faces, these tests are likely to cause at least as much anxiety for parents as for the children taking them.

This isn't a tutorial for how to get your kids through their tests, but it is a chance for you to experience questions similar to the sort of thing that your child will experience. Questions very similar to the ones that follow have appeared in national tests (often referred to as Key Stage 2 'SATs') in the last ten years. If you like, think of our guide to test papers as a Sat-Nav! The questions we've chosen are probably harder than the average test questions, but we chose the questions because of the particular issues that they bring out for parents or children. Have a go at the questions, then read the solutions section, which not only gives the answer, but reveals how parents and children we tested them on fared with these questions, what they had to say about them, and some thoughts on how a mathematician would go about answering them.

Those who are new to these modern tests are often struck

by two things that make them different from the traditional maths test questions of old:

1. They are often more wordy, and it can take a while to work out what the maths question being asked actually is.
2. They often require more than one step to solve them (and they might require knowledge from two different areas of maths, for example shapes and fractions).

The reason for having questions like this is that they are intended to test a child's ability to apply their maths knowledge to general problem solving. In everyday life, situations rarely present themselves as straightforward maths questions – 'What is 2,481 take away 1,923?' – so these tests are part of the preparation for the real world. That, at least, is the theory.

Looking at some questions, you might find yourself asking: 'Where do I even start?' This is a classic situation of Stuck-ness, and there are two tools that you can always use. The first is to ask, 'What DO I know?' The second is to pick one possible answer – even if you're pretty confident it's wrong – and test it. Keep testing like this and patterns will begin to emerge. Many people feel guilty about this 'trial and error' approach, but in fact it's a method used by mathematicians of all abilities. Invariably, clever right solutions are only discovered after seemingly dim, blind-alley approaches have been done first.

Some mums and dads reported being in denial or feeling sick before doing these questions . . . yet when they got down to it, not only was it not as hard, but some reported that they actually found themselves ENJOYING it. What does that say about the maths experience?

PART 1 – CALCULATOR NOT ALLOWED

Q1 7.6 − 2.75 =

Q2 **Karen knows that 74 × 3 = 222**
How does she use this to work out 174 × 3 ?

Q3 **Adam makes this pattern on a grid:**

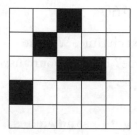

Then he rotates it to a new position. Shade in the missing parts of the pattern:

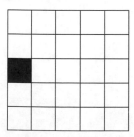

Q4 Peter needs to work out how much two oranges and one apple cost.

Tick all the information that Peter needs to solve his problem:

Oranges cost 10p more than apples ☐

Apples cost 18p ☐

Peter has £1 ☐

Q5 A sequence of numbers begins:

40,   80,   120,   160 . . .

and continues, going up by 40 each time. Will the number 2,140 be in the sequence? Explain your answer.

Q6 For a school outing, everyone gets 3 sandwiches, 2 apples, and 1 packet of crisps.

There are 45 sandwiches altogether. How many crisp packets are there?

Q7 **A spinner is set up on two rectangular boards like this:**

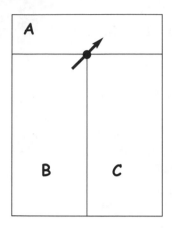
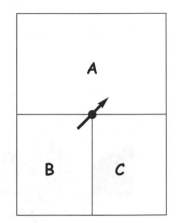

Sam claims that if you spin the spinner on each board, it is more likely to point to region A on the second board than on the first board. Is she right? Explain your answer.

Q8 **How many millilitres of water must be added to this jug in order to fill it up to 400 ml?**

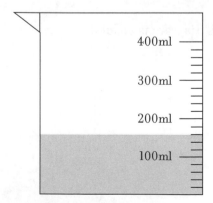

Q9 Alex thinks of a number.

He adds half of the number to a quarter of the number. The result is 60.

What was the number Alex first thought of?

Q10 One third of this square is shaded.

The same square is used in the diagrams below.

a) What fraction of this new diagram is shaded?

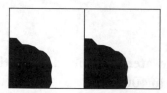

b) What fraction of this diagram is shaded?

Q11 **Here is a number line**.

Estimate where the number 125 is on the line, and mark it with a cross.

0   200

Q12 **A triangle with co-ordinates 1,3 5,3 5,9 is reflected in the dotted line shown.**

One vertex of the reflected triangle is at position 11,3. What are the co-ordinates of the other two vertices of the reflected triangle?

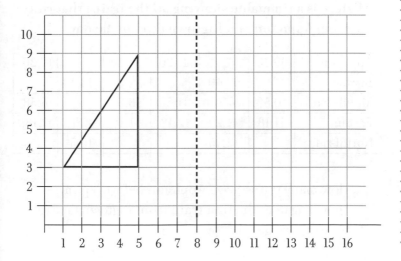

Q13 The shape below is an irregular quadrilateral.
Draw a *rectangle* with the same area on the same grid:

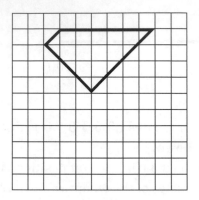

Q14 Here is a timetable showing all the trains that run from Appleton and Bigtown during the day:

Appleton	09.20		11.20	13.20		16.20
Bigtown		09.40			13.40	
Middleton	10.58		11.58	13.58		16.58
Easton	11.18	11.29	12.18	14.18	15.17	17.18
Northbridge		12.05			15.51	

How many trains leave Appleton before 3 p.m.?
How long does the first train from Bigtown take to get to Northbridge?

Q15 **Four small and three large equilateral triangles fit exactly into this rectangle:**

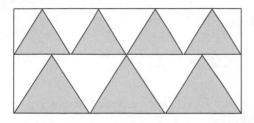

The side of a small triangle is 6 centimetres.
What length is the side of a large triangle?

Q16 **Shade in four sixths of this diagram:**

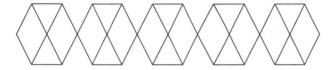

PART 2 – CALCULATORS ARE ALLOWED

Q17 What is 60% of 735?

Q18 What is $12.5 \times (18.9 + 61.1.)$?

Q19 17 multiplied by itself gives a *3-digit* answer:
$17 \times 17 = 289$.
What is the smallest number that when multiplied by itself gives a *4-digit* answer?

Q20 30% of [   ] is 60

Q21

3	3	3
8	8	8

Use five of the number cards above to make this sum correct.

$$
\begin{array}{ccc}
 & \square\square\square \\
+ & \square\square \\
\hline
= 4 & 2 & 6
\end{array}
$$

Q22 Write three prime numbers that multiply to make 385.

$$\square \times \square \times \square \ = \ 3\,8\,5$$

Q23 Here is an equilateral triangle inside a rectangle.
Calculate the value of angle x.

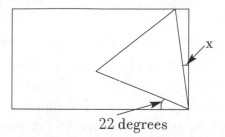

22 degrees

(this is not to scale, so do not use a protractor).

Q24 P and Q stand for two whole numbers.
P is 200 greater than Q. P + Q = 350.
Calculate P and Q.

Q25 Two numbers between 50 and 70 are multiplied together and the answer is 4095. What are the two numbers?

Q26 Here is a cube.
The cube is shaded all the way round so that the top half is shaded and the bottom half is white.

Here is the net of the cube.

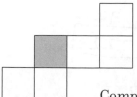

Complete the shading.

ANSWERS

Q1 4.85

Many (but certainly not all) mums and dads find questions like this easy. For children it can be a different story. The confusion comes in the fact that 2.75 has two digits after the decimal point, whereas 7.6 has only one. The sum becomes a bit easier to grasp if you think of the numbers as money: 7.6 becomes £7.60 while 2.75 is £2.75. Now nearly everyone is confident of coming up with the answer, typically by rounding up £2.75 to £3 (that's 25p), then adding £4.60 to make £7.60, so the answer is £4.85. But without that leap to think of it as money – and few children would make that leap – this is an easy question to get wrong without a calculator.

Q2 $3 \times 174 = 3 \times 74 + 3 \times 100 = 222 + 300 = 522$

This is another question that mums and dads tend to get right, though most are happier just doing the calculation rather than explaining *how* they do it. Indeed it's common for adults not to see how the knowledge of 3×74 actually helps here. One mum confronted by this question attempted it by doing the full long multiplication, and found it hard. Her son (aged 8) then showed her how he worked it out using the grid method:

$$= 522$$

Suddenly the mum realised how easy it was and, what's more, now she understood how the sum worked. This is a good example of the 'new' method of doing multiplication helping children (and adults) to understand what is going on.

Q3

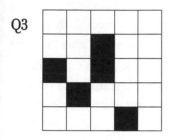

Easy? Most parents think so, but not all. One mum, who had a particular difficulty with the visual side of maths, said that she gave the page a quarter turn so that she could copy the top diagram, only to realise that the bottom of the page had also turned by a quarter. Children are encouraged to use tracing paper for questions like this. This is the first example of a visual/spatial question, and while many adults and children have no problem with this kind of thinking, there are some who really struggle with it. The fact is that some people who are otherwise excellent at mathematical thinking are flummoxed by the visual side of the subject.

Q4 Tick the first two boxes but not the third – the fact that Peter has £1 is irrelevant.

Most parents breeze through this question, though they prefer to just calculate the answer rather than tick the boxes. This is an example of a child being asked to explain their *strategy* for solving a problem rather than actually solving it. It's testing whether the child recognises if a piece of information is redundant or not, an important life skill because in many situations we are presented with lots of information, and need to sift out the relevant parts before we do the maths. However, it's easy to see why a child might get confused by this sort of question. The child has probably had lots of experience of wondering what they can buy with pocket money, and the key question that every child will ask is, 'Have I got enough?' Alternatively, they will ask, 'What exactly CAN I buy, now that I know I have a pound? Instead of buying just two oranges and an apple, maybe I'll have enough to buy some sweets too, so it's really helpful to know how much spare I will have.'

Q5 No, 2,140 is not in the sequence because 2,140 does not divide exactly by 40.

This is typical of the sort of sequence question that children will be set these days. Instead of the familiar IQ question of old that typically asks 'What comes next in this sequence?' the challenge is to see where the sequence is going a long way in the future. So how is it worked out? We found that some mums and dads patiently listed out every number in the sequence, taking several minutes to finally arrive at 2,120, 2,160 . . . and thus demonstrate that 2,140 doesn't appear. Maybe some children do this too, but if so it eats up valuable test time, and they have missed the method that the examiners want them to use.

The most important shortcut is to recognise that all the numbers end in zero, and therefore the numbers can be made simpler by dividing them all by 10. In other words the question is the same as: 'If you keep adding 4, does 214 appear in the sequence?' And this is the same as asking: 'Does 214 divide exactly by 4?' Many mums, dads and children spot this, and proceed to use their own particular technique to do the division.

There is a further shortcut that makes it easier still (though it's not widely known among children or adults). If you want to know whether a number is divisible by 4, you only need to look at *the number formed by the last two digits*. So when checking 214, you just need to look at 14 divided by 4. Now it's easy to see that the number doesn't divide exactly. (Incidentally, one practical use of this technique is that leap years – and Olympic years – always divide exactly by 4. To check if 2012 will be an Olympic year – just in case the news had passed you by – instead of having to divide 2012 by 4, you just need to check if 4 goes into 12 exactly, which it does. So the London organisers can breathe a sigh of relief, they did indeed pick the correct year.)

Q6 There are 15 crisp packets.

There are 45 sandwiches, and three times as many sandwiches as crisps, 45 divided by 3 = 15, so there must be 15 packs of crisps. What could be simpler than that? For most parents this sort of problem is trivial. But that's probably because as adults we do calculations like this as a matter of course, when sorting out catering or doing other everyday chores.

For children, the challenge with this is working out from the word problem what the actual mathematical question is. No mention is made of division in the problem, that's for children to work out. And the best way for them to learn how to do questions like this is through experience. It's a good example of how involving your children in domestic mathematics can help them to recognise what maths is involved almost by instinct.

Incidentally, this question is also interesting for the fact that it includes redundant information. The information about the apples is irrelevant to the problem. Some children will be worried by the apples, as they will assume that they are supposed to use them in some way. Of course real-life problems are full of extraneous information, but trying to simulate real life in maths questions is fraught with difficulty, if the children begin to second guess what the examiner is playing at.

Q7 Sam is wrong, the spinner is equally likely to point to A on both boards.

Here is a question that divides parents (and children) down the middle. For the parents, this is partly because many of them never actually studied probability at school. But the other thing that causes difficulty here is uncertainty over whether in this case the chance of the spinner pointing to a particular region has anything to do with the size of that zone. Because the 'A' zone is bigger on the right-hand diagram, many think that the pointer is more likely to end in this zone. (Otherwise, why did the examiners make this change?) Actually the size of the zones is a red herring. All that matters here is the angle of the spinner. The spinner moves round in a circle, and in both diagrams, half of the circle is in zone A, so the two zones are equivalent. Many children's games involve spinners, so this answer would be easy to verify by testing it out for real.

Q8 240ml of water needs to be added to the jug.
This question is straightforward for most parents, who are quite used to reading scales, but children are often caught out when not all the marks on the scale are numbered. A common error is to think that the current water level is 130 (with each mark indicating 10) instead of 160.

The second step is to work out what the mathematical question is. Here's a great example of how the problem can be treated as a subtraction (400 take away 160), or as an addition (how much do I add to 160 to make 400 . . . first add 40, then add 200). In fact, children and adults are both likely to treat this as an addition, not least because the 'real world' way of tackling this would be to add water to take the level up to 400.

Q9 The original number was 80.
'I don't even know where to begin', said one frustrated parent. To some, this question reads more like a riddle. Many parents recognise that the question could be solved using 'algebra', but since they can't remember how to actually do algebra, they get stuck. Many used trial and error to come up with the answer, but felt guilty about doing so. In fact trial and error − or 'trial and improvement', as it is always referred to in primary schools these days − is exactly how most eleven-year-olds would be expected to solve it. In other words, start by guessing what the number is − try 100, for example − and see what answer that gives you. If the

answer is wrong, adjust the starting number and home in on the right one.

There are plenty of other strategies that might be used. If you would like to rekindle your algebraic skills, here is the classic way of solving the problem. Let's call the number that Alex thinks of 'A' (or, if you're more comfortable using personal names, just call the number Alex). We are told that half of A plus a quarter of A equals 60, or:

$$\tfrac{1}{2} A + \tfrac{1}{4} A = 60$$

Adding the fractions gives you:

$$\tfrac{3}{4} A = 60$$

Divide both sides by $\tfrac{3}{4}$ to give:

$A = 60 \div (\tfrac{3}{4})$ and remembering the rule for dividing by fractions, this means:

$A = 60 \times \tfrac{4}{3}$

So $A = 80$

But remember, it's not expected that 11 year olds will be able to work things out this way – though there are some who are happy and able to do so.

Q10 a) $\tfrac{1}{3}$ b) $\tfrac{1}{9}$

Was your first thought that the answer a) was *two* thirds? If so, you are not alone – the majority of mums and dads we showed this to were 'fooled' at first, before they realised something was wrong – particularly when they looked at the question b) (at which point many parents felt like throwing in the towel).

It's easy to see how children and adults can be confused by a question like this. If you think of each of the shaded squares as a cake, then if you have a third of a cake and add it to another third of a cake, you get two thirds of a cake. But this isn't a question about adding fractions.

This kind of question is really testing the understanding of what fractions mean, but there are many who regard this sort of thing as a deliberate attempt to trick, and therefore unfair. Some adults worked out the first part by saying if this is one third:

then one part is black and two parts are white, so that . . .

is six parts, two of which are black, so the fraction is $\frac{2}{6}$. And $\frac{2}{6}$ is the same as $\frac{1}{3}$, but many adults don't bother to do this simplification (or fail to notice it).

Sometimes it's possible to get the right answer using completely erroneous maths. One dad added $\frac{1}{3} + \frac{1}{3}$ by adding the two numerators $(1 + 1)$ and the two denominators $(3 + 3)$ to get $\frac{2}{6}$ and hence he completely fluked the right answer. If the question had been $\frac{1}{2} + \frac{1}{4}$ he would have added them to get the same answer, $\frac{2}{6}$ which is clearly wrong ($\frac{1}{2} + \frac{1}{4}$ is three quarters, not two sixths!).

For anyone who found the first part of the question hard, the second part came as a cruel twist, and indeed many parents just left part 2 blank. One way to tackle the question is to think of each of the squares as having three equal parts, so there are nine portions in total of which only one has been shaded.

Q11

125 is about here

0   200

Estimation questions can worry children and parents. 'Just how accurate do they expect me to be?' is the most common concern.

In fact, for questions like this, the markers are given a degree of tolerance that they will regard as acceptable, and it's usually quite generous. The main thing they are looking for here is that the child realises that 125 is more than half-way (many children don't), so an arrow pointing to the centre, or to the left half, is clearly wrong. Likewise, an arrow pointing to the final quarter of the line close to 200 also suggests that the child hasn't grasped that 125 is closer to the centre of the line than the end. You can make a very accurate estimate of the position of 125 on the number line by finding the midpoint of the line (100) by eye, then find the point halfway between 100 and 200 (150) and finally find the point halfway between 100 and 150.

Q12 The other vertices are at 11,9 and 15,3.
At least half the horror that parents might have with this question is the use of the word 'vertices'. 'Why can't they just call them corners?' asked one dad. Children are encouraged to learn the mathematical words for things in

their work, and they will be expected to be familiar with terms like 'vertex' and its plural 'vertices'. Parents struggling to remember these terms should take a look at our glossary. Be reassured that after a while, the words become familiar.

The second challenge with this question is that the axes are not labelled, so many mums and dads reported an initial panic as they wondered where position '1,3' is on the diagram. But counting along and comparing the numbers with the co-ordinates you are given makes this obvious (1,3 means 1 along and 3 up). Then it's just a matter of making sure that you don't get sloppy and miscount how far across the right-hand corner (vertex) of the reflected triangle is.

Q13 The area of the quadrilateral is 15 squares, and a 3 × 5 rectangle has the same area.

This question caused significant anguish in many mums and dads. The anguish came in two parts – first in actually working out the area of the quadrilateral, then in converting that into a rectangular area. It's possible to work out the area by dividing up the shape into triangles and working out the area of each triangle. But at this stage in their learning, the main method that children use for working out area is to count squares. The *whole* squares are easy to count, and the remaining squares are all half-squares, making 15 squares in total.

The easiest rectangle to draw that has an area of 15 squares is a 15 × 1 rectangle. But the sneaky examiners have made sure the grid isn't 15 squares wide, so the children now have to work out which other dimensions multiply up to make 15. The only answer is 3 × 5 (or 5 × 3 of course).

Without it being made explicit, the children have been asked to 'factorise' 15, and since 3 and 5 are the *only* factors of 15 apart from 1 and 15 (3 and 5 are prime numbers) there is no other answer to the question.

Q14 Three trains leave before 3p.m. The first train from Bigtown takes 2 hrs 25 minutes to get to Northbridge. A train timetables is just one of many types of table that children will be expected to interpret. Parents who use trains will find this a (relative) doddle, but many children have never taken a train, let alone consulted a timetable, and they might be thrown by the blanks in the table and the convention that train routes are listed in order of times. There's also the challenge of converting between the 24-hour clock and the 12-hour clock (a common mistake being that 13:00 is 3 o'clock, which means many children reckon only two trains leave before this time).

Children (and their parents) will often misread a question, in this case thinking they are looking for the first train, when in fact they are asked to find the first train *that leaves Bigtown* (which is the 9:40 a.m. train). Working out the time difference between 9.40 a.m. and 12.05 p.m. might sound like a subtraction, but attempting to do 'subtraction' on times is far from easy (what is 12:05 take away 9:40 in 'clock' maths?). So the natural, and best, way to work this out is by addition, first adding 20 minutes onto 9:40 to make 10:00, then add 2 hours to make noon, then another 5 minutes, to make a total of 2 hours 25 minutes.

Q15 The larger triangle sides are 8 centimetres.

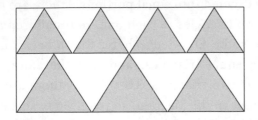

More than one step is needed here. The fact that these are all triangles can be distracting. The only important piece of information is that they are all equilateral, i.e., all three sides are the same. Now it's a matter of saying that 4 × small triangles (6cm) equals 3 × the large triangles. In other words 24 = 3 × the side of a large triangle.

Q16 One of several ways to shade in four sixths is this:

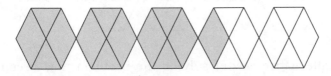

Is this an ingenious brainteaser or a devilishly mean trick question? It depends on who you ask (though more people belong in the second camp than the first). For many adults and children there is a heart-sinking moment when you realise that you are being asked to divide *five* objects into *sixths*. Children will often declare to their teacher at this point that the question is 'impossible' and will make no further attempt. Even reasoning with them that the examiners wouldn't have set a question if it were impossible cuts little ice.

What adds to the difficulty of this question is that the hexagons aren't divided into equal portions – there are two diamonds and two triangles in each one. Your children are expected to use some knowledge here, that a hexagon is made up of six equal triangles, and that therefore the diamond must be two triangles. That means there are 30 triangles altogether in the diagram, so you need to shade in four sixths (or 20) of them. There are various ways of doing it, including the one shown above. Another way is simply to shade in all of the diamonds.

The problem is much easier to solve visually if you draw in the horizontal line through the centre of the hexagons. All is now much clearer – but most parents and almost all children don't think of this.

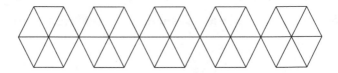

An alternative way of solving the problem now becomes more obvious – that you can simply shade in four out of every six of the triangles in each hexagon:

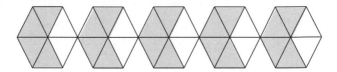

Q17 60% of 735 is 441.
Calculators can be a curse sometimes, especially those that have prominent percentage buttons. The temptation is to

do a sum like this one by typing the buttons ⬚6⬚ ⬚0⬚ ⬚%⬚ , at which point it suddenly dawns on many adults that they can't remember how to use the percentage button.

Much better is to use a bit of common sense to think about roughly how big the answer is going to be. 60% is not that far from 50%, which is a half, so the question is asking you to find roughly half of 735. In other words, the answer is going to be not that far from 400.

Children learn some interesting strategies for this. One is to work out 10% of the answer, which is 73.5, and then multiply that by 6 to give them 60%.

But the quickest route is to remind yourself that 'percent of' means 'divided by one hundred, multiplied by . . .'. In other words, 60% of 735 can be rewritten as $60 \div 100 \times 735$. ($735 \times 60 \div 100$ gives the same answer.)

Q18 1,000

Calculators are only as good as the person operating them. You probably know instinctively that you should add 18.9. and 61.1 before multiplying by 12.5 (this is BODMAS, see page 281), but it's not so automatic for your children, who will be inclined to work from the left, and without brackets on their calculator will probably end up with the answer 297.35.

Q19 The smallest number is 32 (32 × 32 = 1,024).

The maths involved isn't hard here, but children are expected to grasp ideas about 'four-digit numbers' and many struggle with the language and the rather abstract nature of the question. Once again, this is a problem easily solved with trial and improvement, so $30 \times 30 = 900$ (too small), $35 \times 35 = 1,225$ (too big) and the right answer is found by searching between those numbers.

All this trial and improvement is crying out for a short-cut, and there is one for children who are confident with the idea that multiplying a number by itself is called 'squaring' the number. To find the number that when squared produces the smallest four-digit number, simply enter 1,000 into your calculator and then press the square root button $\boxed{\surd}$. The square root of 1,000 is 31.62 . . . The first whole number that produces a four-digit number when squared is simply found by rounding up 31.62 to the next whole number, which is 32.

Q20 200 (in other words, 30% of 200 is 60).

Now if only this question had been 'what is 30% of 200?' then it would have been as easy as question 11. But instead, those sneaky examiners have turned the question round. And children can really struggle with these questions when they aren't being asked the question in the normal way. For many, the only strategy once again will be by trial and improvement, and indeed many parents tackled it in just the same way. But there are lots of other strategies. Here are two:

- Start by realising that if 30% of the number is 60, then 10% of the number must be 20 (you've just divided everything by 3). And if 10% is 20, then 100% is 200.

- If 30% of X is 60, then X must be 60 divided by 30%, which is the same as calculating $60 \div 0.3$, which (your calculator will confirm) is 200.

Q21

```
      3 8              8 8
  + 3 8 8    or    3 3 8
```

This is a question with more than one answer, though this is more likely to worry parents than children (who are generally happy to be able to find any answer at all). Creating a sum rather than carrying out a set of rules for doing a sum is much more of a challenge, and once more the expectation is that the children should just have a go before finding a strategy for solving the problem. The most obvious clue in this particular mystery is that if the numbers add to 426, then the hundreds digit in the second number cannot possibly be 8, so it must be 3. The other main clue is that the two units numbers added together end in 6. This could be 3 + 3 or 8 + 8. The 'tens' numbers must be 3 and 8, and further trial leads to the answers above.

Q22 The prime numbers are 5, 7 and 11.

This is a 'detective' question, which this time includes the need to remember what *prime number* means – easy enough when you encounter them every day, but easy to forget if you haven't touched prime numbers since your schooldays. The question could have excluded the word 'prime' and asked just to find any three numbers that multiply to make 385, but no doubt many would have opted for $1 \times 1 \times 385$ which side-steps what the examiners are looking for.

So instead we are looking for three numbers, none of which is 1, that multiply to make 385 (and it so happens that they are prime numbers). The immediate clue here is the 5 at the end of 385, which means that 5 must be one of the divisors. Now divide 385 by 5 to get 77, and of course $77 = 7 \times 11$.

Q23 The angle X is 8 degrees.

Many parents throw their hands up at this question, as it requires knowledge that – for most mums and dads – is a distant memory. Two key things to know are that a rectangle has right angles at each corner, each of which are 90 degrees, while an equilateral triangle has three identical 60 degree angles at the corners. 'X' is then calculated by creating the sum: $X = 90 - 60 - 22$.

If you're worried that your child would struggle to work out the answer here with all these steps, then you might be reassured to learn that less than 20 per cent of eleven-year-olds would get the right answer here – it's definitely at the difficult end of what they have to deal with. Some children are particularly intimidated by using a letter like 'X' to label the unknown angle. You can make the problem much friendlier by using our naming idea, calling 'X' something unusual or more human, like 'Blob' or 'Beryl'.

Q24 P is 275, Q is 75.

'Wow – this looks like algebra,' said one dad. Certainly it's an early exposure of children to something that, in later years, they will solve using 'simultaneous equations'. Many children will find it frustrating that they can find an example where P is 200 more than Q, and one where $P + Q$ add to 350, but they can't find both. As ever, this one is down to our old friend trial and improvement. Try $P = 200$, $Q = 0$ and the sum is 200, which is much too small. So try $P = 300$, $Q = 100$, and you end with 400, a bit too big. Further jiggling up and down delivers the correct answer of 275 and 75. (It might seem surprising that these numbers end in 5, when the numbers in the question all ended with zero.)

Q25 The answer is 63 × 65.

Of all the questions in this test, this one produced some of the most anguished reactions. One mother spoke for many others when she said, 'I feel like an imbecile! At first glance I just thought, "No, I can't do that".' And yet, like most mums and dads, she did get the right answer in the end. Though it took her a while.

As in so many examples, the strategy used by parents is the one that is also expected of children for tackling this question, namely trial and improvement. This can mean starting by picking any two numbers at random, 58 × 67 for example, and seeing if they come close to the answer. But without being methodical, this approach can be time-consuming and distressing, as you find yourself shooting above and below the right answer each time.

Solving maths questions can be like cracking a code, and this is a perfect example. The question is a mystery, and like the detective Colombo, you have been presented with the evidence and you are now looking for clues. And, believe it or not, there are plenty of clues. The first is that the number 4,095 is odd. If you multiply two numbers together and the answer is odd, then the two numbers themselves must also be odd. (This was a revelation to one bright primary head teacher we spoke to – she'd never come across the fact that an even number multiplied by any other whole number must result in an even number, though it was obvious to her when she thought about it.)

So immediately we know that the two numbers involved must come from the collection 51, 53, 55, 57, 59, 61, 63, 65, 67 and 69. But the options are reduced further when you realise that if the product ends in a 5, then at least one of the numbers ends in a 5 too. So one of the numbers must be 55 or 65. A quick check on the calculator reveals that 4,095 does not divide exactly by 55, while 4,095 ÷ 65 = 63, revealing the second number.

In fact, mathematicians well practised in this sort of sum would be able to answer the question without needing a calculator at all. If you want to know how (and perhaps you don't!), then read on. First, a mathematician would look at the *factors* of 4,095. They would realise that because 4,095 is odd, both its divisors must be odd. They would spot instantly that 4,095 is divisible by 5 so one of the divisors of 4,095 must be 55 or 65. They would then spot that 4,095 isn't divisible by 11, so that 55 can't be one of the divisors, leaving 65 as the only option. Finally, they would note that 4,095 is divisible by 9, so the other divisor must be an odd multiple of 9 between 50 and 70, which means it must be 63.

Don't worry if that quick explanation left you none the wiser, the main point is that with enough maths knowledge you can learn shortcuts that will enable you to crack seemingly tough questions very quickly.

None of this is *required* of an eleven-year-old. And yet . . . many children are more than capable of understanding these shortcuts, and by encouraging them to explore number mysteries like this, not only are you helping them to solve questions faster, you are also helping them to discover that maths isn't just a matter of blind trial and error, but can be more a case of detective work and discovery. And that makes it a far more interesting subject.

Q26

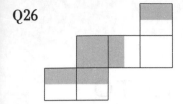

'If there's one question that strikes terror into me,' said one primary teacher, 'it's "net" questions. They are my absolute weak spot, my Achilles heel.' The ability to mentally fold up a flat piece of card to form a simple 3-dimensional object (and to imagine the opposite – what a solid object looks like when unfolded to make a flat net) is expected of the modern Year 6 child. And it's a challenge that many children find very, very difficult. To understand what is going on, most children need actually to draw out the net and then physically make it into the object. At that point they will experience the aha! moment. Unfortunately, making a model takes time, and only the most adventurous and quick-working of children would be able to do this in less than five minutes – so it would eat into valuable exam time. The only way round this is for children to have plenty of practice making nets outside exam time and to picture folding their nets before they actually do so.

ANSWERS TO TEST YOURSELF QUESTIONS

Numbers and place value

i) MDCLXVI is the number 1666, the year of the Great Fire of London. You'll find it inscribed on 'The Monument', close to London Bridge.

ii) It helps to think of the number split up into its component parts. 124 can be written as $100 + 20 + 4$. Remember that we are working in base 8 so 20 does not stand for twenty (that is 2 groups of 10) it stands for 2 groups of 8, so that's 16 in base 10. 100 in base 8 stands for 1 group of *8 groups of 8* – 64 in base 10. So 124 in base 8 is $64 + 16 + 4$ or 84 in base ten. Confused? Image how your child feels getting to grips with place value for the first time.

Addition and subtraction: mental methods
i) *Mental or pencil and paper?*

Most of the sums lend themselves to mental arithmetic.
a. Mental: two from 152 and added to 148 turns this into $150 + 150 = 300$.

b. Mental: 150 from 300 is 150, so 148 from 300 must be 152.

c. Messy numbers – paper and pencil or calculator probably sensible.

d. Mental: Although very similar looking to c., 698 is close to 700 and $843 - 700 = 143$, that's taking away 2 too many, 145.

e. Mental: 3 off the 5,003 and added onto 4,997 makes this $5,000 + 5,000$.

f. Mental: 2,003 (6,002 subtract 4,000 is 2,002, adjust the answer to one more as 4,000 is taking away one extra).

ii) *Number line*

$$48 + 36 = 84$$

iii) *More number lines*

$$73 - 28 = 45$$

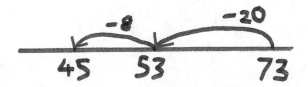

iv) *Shoe prices*

a. To work out the change from £50, it's easier to calculate
by adding:

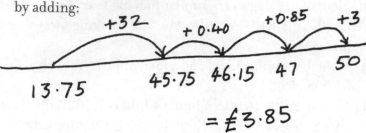

= £3.85

b. For finding the price difference, you probably used
more conventional subtraction, with compensation, for
example: £32.40 − £13.75 = £32.40 − £14 + £0.25 =
£18.65.

Addition and subtraction: paper and pencil methods

i) *Addition using 'partitioning'*

a. 147 + 242

100	40	7
200	40	2
300	80	9

= 389

b. $368 + 772 =$

300	60	8
700	70	2
1000	130	10

$= 1140$

ii) *Subtraction using 'partitioning'*

a. $847 - 623 =$

800	40	7
600	20	3
200	20	4

$= 224$

b. Here's one way:

$721 - 184 =$

600	110	11
100	80	4
500	30	7

$= 537$

iii) *Why must these answers be wrong?*

a. $3,865 + 2,897 = 6,761$ . . . The last digit must be a 2 because $5 + 7$ ends in 2. (Another way of instantly spotting an error is that when two odd numbers are added together, the answer is always even.)

   b. $4,705 + 3,797 = 9,502$ . . . Since 4,705 is less than 5,000
     and 3,797 is less than 4,000, the answer must be less than
     $5,000 + 4,000$.

   c. $3,798 - 2,897 = 1,091$ . . . The answer must be less than
     $3,800 - 2,900$, so it must be less than 1,000.

Simple multiplication and tables

i) *Working out 8 x 7*

8×7 is the same as . . .
$2 \times 7 = 14$
Doubled to make 28
Doubled again 56

ii) *Compensation method*

9×78 is the same as 10×78 ($= 780$) take away 78, which
702. (You could also get there using a different way using
compensation. Work out 9×80 ($= 720$) and subtracting
9×2 ($= 18$) to give the same answer).

iii) *Fairy cakes*

The calculation is 60×9. This is the same as 6×9 ($= 54$)
multiplied by 10 ($= 540$). Did you use the 6 times table or
9 times table to work it out (or couldn't you tell!)?

iv) *Match up the sums*

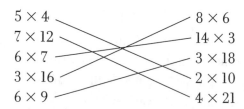

5×4	8×6
7×12	14×3
6×7	3×18
3×16	2×10
6×9	4×21

v) *Elevens*

a. $33 \times 11 = 363$
b. $11 \times 62 = 682$
c. $47 \times 11 = 517$

Multiplication beyond tables

i) *Grid method 1*

$$200 + 60 + 30 + 9 = 299$$

ii) *Why must these answers be wrong?*

1. $37 \times 46 = 1,831$ ... The last digit must be even because 46 is even.
2. $72 \times 31 = 2,072$ ... $70 \times 30 = 2,100$, so the answer must be bigger 2100.
3. $847 \times 92 = 102,714$ ... $1,000 \times 100 = 100,000$ so the answer must be smaller than 100,000.

iii) *Grid method 2*

There's a decimal point — but don't panic. £9.47 is the same as 947 pence. The calculation is therefore 947×62 pence, which you could do like this:

	900	40	7
60	54,000	2,400	420
2	1800	80	14

And if you add that lot up it comes to 58,714 pence, or £587.14 (As you see, for large calculations like this, the grid method becomes cumbersome – but it does work!)

Division

i) *Number sequence*

$$43 \quad 34 \quad 25 \quad 16 \quad 7$$

This is a good example of how division and subtraction can be linked. First, notice that the gap between the start and finish is $43 - 7 = 36$, and that it takes four steps to get from 43 down to 7. So each step must be $36 \div 4 = 9$, making the gap between each of the numbers in the sequence 9.

ii) *Spot the primes*

37, 47 and 67 are prime numbers.

iii) *Factor sorting*

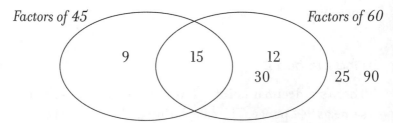

iv) *Divisibility tests*

a. 28,734 ÷ 2   Yes (last digit is even)
b. 9,817 ÷ 5   No (last digit is not 0 or 5)
c. 183 ÷ 3   Yes (digits add to 12)
d. 4,837 ÷ 9   No (digits add to 22)
e. 2,8316 ÷ 6   No (digits add to 20)

v) *Chunking 1*

$$
\begin{array}{r}
8\overline{)336} \\
320 \quad 40\times \\
\hline
16 \\
16 \quad 2\times \\
\hline
= 42
\end{array}
$$

vi) *Chunking 2*

$$
\begin{array}{r}
22\overline{)739} \\
660 \quad 30\times \\
\hline
79 \\
66 \quad 3\times \\
\hline
13 \\
= 33 \, r13
\end{array}
$$

vii) *Why must these answers be wrong?*

a. 223 ÷ 3 = 71 . . . 223 is not exactly divisible by 3 (its digits add to 7) so the answer cannot be a whole number.

b. 71.8 ÷ 8.1 = 9.12 . . . This calculation is close to 72 ÷ 8 (= 9), and the answer must be smaller than 9 because 71.8 is smaller than 72 and 8.1 is larger than 8.

c. 161.483 ÷ 40.32 = 41.36 . . . Don't be distracted by all those decimal places. The approximate answer to this sum is going to be roughly 160 ÷ 40 = 4. So the answer is wrong by a factor of about ten!

Fractions, percentages and decimals

i) *Sausage fractions*

a. $\frac{6}{7}$ is larger — more sausages for the same number of children.

b. $\frac{4}{11}$ is larger — more sausages AND fewer children

c. Can't tell using sausage reasoning

ii) *A huge fraction to simplify*

The 48 can be eliminated by dividing it by the 6 × 4 × 2 on the bottom line:

$$\frac{49 \times \cancel{48} \times 47 \times 46 \times 45 \times 44}{\cancel{6} \times 5 \times \cancel{4} \times 3 \times \cancel{2} \times 1}$$

Next the 45 can be divided by 5 × 3 (15) to leave 3.

$$\frac{49 \times 47 \times 46 \times \cancel{45}^{3} \times 44}{\cancel{6} \times \cancel{5} \times 1}$$

So now the sum becomes $49 \times 47 \times 46 \times 3 \times 44$. A calculator will confirm that this works out at roughly 14 million.

This fraction represents the number of different combinations that you can choose for entering the main UK National Lottery. The top line ($49 \times 48 \times 47 \times 46 \times 45 \times 44$) is the number of different ways that balls can emerge from the lottery machine – the first ball can be any of 49 numbers, leaving 48 for the second, 47 for the third and so on. However, the order in which the balls are drawn does not matter, and $6 \times 5 \times 4 \times 3 \times 2 \times 1$ is the number of ways in which your six chosen numbers could emerge.

What the result means is that on average, every fourteen million times you played the lottery, you would win the jackpot once. Don't give up the day job!

iii) *Chocolate bar fractions*

a. Create a chocolate bar with 3 rows and 11 columns, which means there are 33 chunks. $\frac{2}{3} = \frac{22}{33}$ and $\frac{7}{11} = \frac{21}{33}$ so $\frac{2}{3}$ is larger.
b. Since $\frac{2}{3}$ is $\frac{22}{33}$ and $\frac{7}{11}$ is $\frac{21}{33}$, their sum is $(22 + 21) \div 33$, or $\frac{43}{33}$.

iv) *The wise man and the camel*

The secret to this ancient mystery lies in the fractions of a camel that each son was allocated. The easy way of dividing the camels would have been to give each son one third, and of course $\frac{1}{3} + \frac{1}{3} + \frac{1}{3} = 1$. The father could also have divided the camels up as $\frac{1}{2} + \frac{1}{4} + \frac{1}{4}$, or in lots of other ways, but whichever way he divided them up, the fractions should add up to 1.

Let's look at what the father actually did:

First son: $\frac{1}{2}$ Second son: $\frac{1}{3}$ Third son: $\frac{1}{9}$

To add up these fractions, we need to give them a common denominator, in this case 18. Expressed as 18ths the sons get the following fractions:

First son: $\frac{9}{18}$ Second son: $\frac{6}{18}$ Third son: $\frac{2}{18}$

Add those up and you'll notice something peculiar: $9 + 6 + 2 = 17$, so the father is only giving away $\frac{17}{18}$ths of his camels, not all of them! And $\frac{17}{18}$ths of 17 is messy, it's actually $16\frac{1}{18}$, with $\frac{17}{18}$ths of a camel left over.

When the wise man lends his camel, there are now 18, which means that the sons will be receiving $\frac{17}{18}$ths of 18 – in other words, the sons will now get all of the 17 camels. The wise man can now take away his camel leaving everyone happy.

v) *Percentages*

a. 33 out of 220 is the same as 3 out of 20, or 15%.

b. 40% of £45 is the same as 0.4 × 45 or £18. So the sales price is $45 - 18 = 27$.

c. Most adults instinctively think that the better deal is to add the VAT first and then take 10% off the higher price, since this gives a 'bigger discount'. The correct answer, however, is that *it makes no difference which way you do it!* Although it may not be immediately obvious, if you think about it, making a 10% discount on a price is the same as multiplying the price by 0.9. Adding 20% VAT is the same as multiplying the price by 1.2 This is an example of how the order of multiplication doesn't matter (see page 125). Price × 0.9 × 1.2 is the same as Price × 1.2 × 0.9. OK, that's what the maths says, but if you find it befuddling you are not alone. In our experience many adults still need to check it out on a calculator before they are convinced.

Shapes, symmetry and angles

i) *The tiled floor*

You only need three colours of tile. We've labelled the three colours A, B and C:

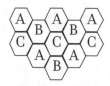

Notice how there are regular patterns of ABC, ACB, and so on in every direction.

ii) *A strange net*

The folded shape forms a 'triangular prism' like a piece of cheese:

iii) *The right-angled triangle*

The triangle could be an isosceles

or scalene

iv) *Parked car*

The corner of the car is 90 degrees and the three angles add to 180 degrees so: A = 180 − 90 − 65, or 25 degrees.

v) *Cocktail sticks*

Remove the 'back leg' to give you this symmetrical wine-glass shape:

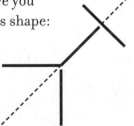

vi) *Where is the square?*

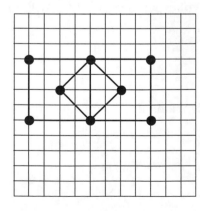

(1, 5) and (1, 9)
(9, 5) and (9, 9)
(3, 7) and (7, 7) (a lot of
people miss this one!)

Measuring

i) *Clock puzzle*

The hands of the clock are at right angles at 3 o'clock and 9 o'clock – those are the two that are easiest to spot. Children might also think of quarter past 12, but be carcful, this isn't quite a right angle because the hour hand is not pointing straight up, it is a quarter of the way between 12 and 1. But there will be a time soon afterwards – just before 12.17, in fact – when the hands are at right angles. And this will occur again just after 1.20, 2.25 and so on. Count the times up carefully, and you should find 11 times when the minute hand is a quarter turn ahead of the hour hand, and eleven times when it is behind, making **twenty-two** times altogether.

ii) *Baking the cake*

The cake comes out at 6.10 p.m. (Recognising that 90 minutes is one and a half hours makes this easier to work out.)

iii) *Time line*

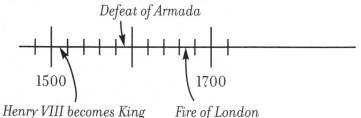

Defeat of Armada

1500

1700

Henry VIII becomes King   *Fire of London*

iv) *Perimeter*

First work out that the side of each square
is 5cm. Counting then establishes
that the perimeter is 14 edges of a
square or 70cm.

v) *The Mysterious appearing square*

If you look carefully, the diagonal line that runs from
piece A to piece D in the rectangle is not a straight line, it
has a slight kink in it. In fact, the diagonal is really a long
thin parallelogram whose area is exactly one unit.

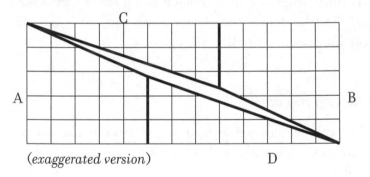

A

C

B

D

(*exaggerated version*)

Data-handling and chance

i) *Carroll diagram*

The Carroll
diagram looks
like this:

	MULTIPLE OF 2	NOT A MULTIPLE OF 2
MULTIPLE OF 5	10 20 30	5 15 25
NOT A MULTIPLE OF 5	4 6 12 18 22	7 11 19

ii) *Which sport?*

100m sprint; the runner starts slowly, rapidly builds up speed, reaches and stays at a maximum speed, crosses the winning line, very rapidly slows down and walks.

iii) *Pie chart*

a. 25 per cent chose computer.
b. 8 children chose sports. (If 10 children chose computer we know that is 25 per cent of the total. Reading and sports together make up 25 per cent, which is therefore also 10 children, so if 2 children chose reading, 8 must have chosen sports.)
c. 25 per cent is 10 children, so 100 per cent must be 40 children.

iv) Dice scores

Arrange the scores in order: 2, 4, 5, 6, 6, 7, 8, 9, 11, 11. The median is 6.5 (midway between 6 and 7), the mean is 6.9 (69 ÷ 10) and there are actually *two* modes, 6 and 11, because each appears twice.

Calculator maths
i) *BODMAS*

a. 16
b. 161
c. 22

ii) *Consecutive numbers*

Use a calculator to find the square root of 4,692, which is roughly 68.4. The two consecutive numbers must be either side of 68.4, in other words 68 and 69. Multiply them together to confirm they give the answer 4,692.

AND
FINALLY . . .

DOs and DON'Ts

If you want your child to enjoy their maths and do well at it (and those two things *do* usually go together) then what you do as a mum or dad will have a huge influence. We've summed these up as a series of Dos and Don'ts. Of course, every child is different, and what works in one family doesn't always work in another.

DOs

Play (maths) with your child

If we had to limit ourselves to one 'Do', it would be this. Games are full of maths, and are the ideal way to engage a child in mathematical thinking. Parents don't play games with their children as much as they used to. Partly it's a matter of time – we're all so much busier these days. But also, so many exciting electronic games are now available, that children are naturally inclined to log on to CBeebies or sit in front of a Gameboy to play. And if the activities they are playing involve maths, then isn't this a win for everybody? After all, if the child is fully engaged and learning maths, the parent can go off and do something else without feeling guilty.

Unfortunately, what's missing is the impromptu learning that happens when a parent and child are playing together. As the child is about to roll the dice, Mum can say, 'Uh-oh – how many do you have to roll if you want to land on me?' In Monopoly you might say, 'Can you be the banker and change my £500 into hundreds and fifties?' The board games, car games and other interactive games that we have described throughout this book are the perfect way of naturally introducing mathematical ideas to your child.

Let your child win, or be 'better than you'

If you always win or get the right answer, there's a danger that all your child will learn is that you are good at maths. And children are like the rest of us: if we keep losing every time, we begin to think there must be better games to play then this. Of course you know your child better than anyone, so you'll know what the right balance is for how often you should let them win. Letting your child win *every* time doesn't prepare them for failure – and nobody likes a spoilt brat!

Incidentally, you can introduce mathematical ideas even when you're letting your child 'win' at something that's nothing to do with maths. For example, most parents are familiar with the challenge of getting the children ready for bed. A challenge to beat the clock is usually an incentive that children enjoy: 'I bet you won't be ready by the time I've counted to . . . thirteen.' You then pace your counting to make sure it's going to be a tight finish, which gives you a great chance to end up doing something like this: '. . . eleven, eleven and a half, twelve, twelve and a half, twelve and three quarters, twelve and seven eighths . . .' thus subliminally introducing your child to the notion of increasing fractions . . . Oh, what subversive people we are ☺.

Make maths a casual part of what you do while you're doing something else

How does it feel to have Mum or Dad sit down with you at the table and say, 'Right, let's do some sums'? For many children it is the same feeling as when Mum or Dad piles that squishy carrot onto the fork and pushes it across at you saying, 'Eat up, you know this is good for you.' The immediate reaction is to resist. Instead of making maths a formal thing to do with your child, find ways of sneaking it in casually. As you unload the dishwasher, say, 'Oh, I hope there are going to be enough bowls for dinner tonight – let's check.' As you pack the shopping, ask, 'If the bread is 84 pence and the milk is 33 pence, what does that come to?' (You don't need to make it clear if you are asking yourself or your child this question.) While you are waiting in the checkout queue try to estimate how much the shopping will come to. Is £1 an item a good estimate? Who got closer to the actual amount? Or even, as you are walking to school and a Number 23 bus goes past, suddenly ask, 'Hey, I was just thinking, is 23 a prime number?' Introducing a maths idea like this without any apparent reason gives out the message that maths is something that you might randomly talk about just like the weather or any other everyday topic.

Do lots of hands-on maths

Remember the three Cs of everyday maths: 'Cash, clocks and cooking'. All three are perfect opportunities to practise maths. Instead of paying for the shopping, give your child the chance to do it. Get them to help you to measure ingredients. Give them a watch to wear so that they can help you to tell the time. There are lots of ways to do this in a safe, unthreatening manner. For example, one nice tactic when telling the time is to say to your child (casually, as ever), 'I make it half past eleven, is that what you make it?' And cooking is the ideal place for introducing all sorts of maths, including measuring, converting from imperial to metric, fractions ('half a pound'), ratios ('this recipe is for four people, we need to work out the ingredients for ten people') and multiplication ('this cake tin has three rows and four columns, how many fairy cakes will it hold?').

Make it silly, gruesome, scary or dangerous

How can you get a child excited about answering a question like 'what is 7 plus 11'? The answer for most children is to make the outcome exciting. The simple challenge of 'I bet you don't know what 7 + 11 is' is enough to spur many children on. But you can be more adventurous. One dad we know used to huddle his children together and whisper, in a conspiratorial way, things like, 'Hey, let's sneak over to Mr Pumfrey's house and *write the eight times table* in chalk on his path.' Presented as an adventure, the children were only too delighted to take part. Oh how they'd chuckle as they scribbled $4 \times 8 = 32$, and dashed away before Mr Pumfrey (who was of course completely in on the deal) could come out and catch them in the act. Another sure-fire technique is to announce you will do something extremely juvenile, such as let off a huge raspberry, if anyone comes up with the right answer to the particular maths question that you want your child to tackle. With that incentive, there aren't many children below the age of eight who won't keep trying.

Recognise that there's more than one way of doing calculations

Children will discover their own methods of solving maths problems — sometimes these are long-winded, sometimes they are shortcuts. You can guide them towards methods that are quick and which you know work, but you can't — and shouldn't — force a child to use a method that means nothing to them. And in any case, there isn't one single method that is the best method for all problems. For example $3,786 + 4,999$ is sensibly calculated using a mental arithmetic method of adding $3,785 + 5,000$, whereas $3,786 + 4,568$ needs paper and pencil or a calculator. 45×99 is easy to calculate mentally ($45 \times 100 = 4,500$, that's 45 too many, 4.455) but 45×68 is not.

Be a geek

As sophisticated adults, we all 'know' what the 'geeky' or 'sad' things are in life: trainspotting, stamp collecting, searching for prime numbers, those sorts of things. Our culture reminds us of just how geeky these things are all the time. Except, of course, the truth is that most of us are secret geeks, and the biggest geeks of all are children. Most children love obsessive, repetitive tasks, abstract games and getting excited about what adults might regard as banal topics. The notion of geekiness or activities being 'sad' is not one that most young children identify with, and in a way the only sad thing is that these notions get imposed on them. If you announce, 'Hey children, let's check what the number of this train is,' then chances are they will find this exciting – as long as you are able to sound excited about it yourself. Yes, your children are trainspotters. Get over it.

Learn to be an actor

When you say 'well done, the effort that you put into getting that answer was fantastic', say it like you mean it. Get excited about mathematical ideas, and your children will get excited too. Who knows, if you fake it for long enough, you might actually start genuinely getting excited yourself – it can be infectious.

DON'Ts

Our list of don'ts is very short. In fact, we only have two items in the list. But they are very important ones:

Don't expect them to 'get it' after you've explained it once

Don't even expect them to get it after you've told them fifty times. It can take a long, long time for the penny to drop and for mathematical ideas to become second nature. A child will know that seven sevens are 49 one day, but then the next day, when it crops up in a different context, suddenly they think the answer is 47. *This is normal.* Remember, it probably took you years to get to your current level of competence with numbers (whatever that level might be).

Don't tell them that you are hopeless at maths

And in particular, don't *revel* in the fact. It's common for adults to say, 'I was always hopeless at maths', almost as a boast. Why is this? Partly it's because many adults believe that this is true, because they so vividly remember finding maths hard, and can remember that their maths lessons were filled with times when they got things wrong. Partly it's a defence mechanism: by claiming to be hopeless at maths, they know you won't ask them a maths question. But the claim to be hopeless at maths can also carry a more insidious message: '. . . and look, I'm a successful adult now, so clearly being good at maths is not that important.'

Children pick up on these messages, and the main effect is to switch them off maths. It builds up an expectation that maths is going to be something that they won't enjoy, that will be filled with failure, and ultimately won't be any use to them anyway.

The truth is that many parents who claim to be hopeless at maths are being disingenuous. Those same parents often have no difficulty managing household budgets, working out itineraries, juggling multiple tasks and playing strategy games.

It's often said that people that boast about being hopeless at maths never make the same boast about being hopeless at spelling or reading. This might be true in some circumstances, but part of the reason why it happens is that people confuse the broad subject 'maths' with the narrow field of arithmetic. If somebody introduced themselves as an English teacher, you wouldn't immediately back off and say 'gosh, I was always hopeless at spelling' because you realise that English is about ideas and imagination, not just the nitty gritty of grammar, spelling and punctuation. Yet that is the equivalent of what people do with maths.

School experience, reinforced by a popular cultural prejudice, have reduced our perception of maths to being little more than difficult arithmetic. The truth is that maths is far, far more than that. It is a creative, imaginative, and deeply philosophical subject. Unfortunately, the pressures of the curriculum, and having to deal with a large group of children with a wide range of abilities, can limit the level to which teachers are able to present this creative side to the subject (though there are many teachers who, against the odds, manage to do this).

Fortunately, there is somewhere else where your children can learn more about what real maths is, the maths that they can enjoy and that is imaginative. And that's with you, Mum and Dad.

ACKNOWLEDGMENTS

This book would not have got off the ground without the help of the many parents, teachers and children who gave us their valuable perspectives on what maths means to them, and where they find it challenging. Thanks go to the teachers and parents at St Catherine's School in Bletchingley, Heber School in East Dulwich and Andrew's Endowed School in Holybourne who were so willing to share their personal experiences. Thanks to the collective primary teachers of Liverpool who gave us so many wonderful and heart-warming anecdotes. A very special thank you to Stephanie Gibson, Emily Joyce and Elaine Standish, who allowed us to test so much material on them, often on weekday evenings when they had every right to be doing something more relaxing than a SAT question.

Thanks also to Ted Dexter, Rivka Rosenberg, Mike Tuer, Andrew Mylius, Helen Lowe, Helen Porter, Jenny Jones, Phil Rees, Charlotte Howard, Chelsey Fox, Sean Flynn, Andrew Cunningham, Mandy Freshwater, Boo, Patricia Reid and anyone else that we've inadvertently forgotten to mention.

Of the children who helped us, William, Georgia and Jenna deserve special mention for some immaculate work, and there are plenty of others whom we shall leave anonymous, but who gave us plenty of food for thought (and lots to smile at, too.)

Some of the examples we've used come directly from the work of leading experts in the field. We'd like to acknowledge:

Julia Anghileri of Cambridge University for the example of 6000 ÷ 6 on page 157.

Martin Hughes of Exeter University for the story of the child saying 'I don't go to school yet' on page 88.

Malcolm Swann of Nottingham University for the graph examples on page 258.

Nan Flowerdew for 'Be kind to pepol' on page 272.

And finally thank you to our wonderful editor Rosemary Davidson, who wore her heart on her sleeve to ensure that we were addressing the real concerns of mums and dads, and to Peter Ward for services to typesetting well above and beyond the call of duty.